Trends in Maritime Autonomous Surface Ships (MASS) Technology

Online at: https://doi.org/10.1088/978-0-7503-3151-7

Trends in Maritime Autonomous Surface Ships (MASS) Technology

Hsin-Hung Cheng
Institute of Transportation, Ministry of Transportation and Communication, Taiwan

IOP Publishing, Bristol, UK

ISBN 978-0-7503-3151-7 (ebook)
ISBN 978-0-7503-3149-4 (print)
ISBN 978-0-7503-3152-4 (myPrint)
ISBN 978-0-7503-3150-0 (mobi)

DOI 10.1088/978-0-7503-3151-7

Version: 20231001

IOP ebooks

British Library Cataloguing-in-Publication Data: A catalogue record for this book is available from the British Library.

Published by IOP Publishing, wholly owned by The Institute of Physics, London

IOP Publishing, No.2 The Distillery, Glassfields, Avon Street, Bristol, BS2 0GR, UK

US Office: IOP Publishing, Inc., 190 North Independence Mall West, Suite 601, Philadelphia, PA 19106, USA

To my family.

Contents

Preface ix

Author biography xi

1 Introduction 1-1

1.1 The history of ships 1-1
1.2 Review of the development of ships 1-2
1.2.1 Raft ships 1-2
1.2.2 Sailing ships 1-3
1.2.3 Steamships 1-4
1.2.4 Internal combustion engine (ICE) ship 1-5
1.3 Future trends 1-6
References 1-7

2 Background of artificial intelligence (AI) 2-1

2.1 Review of AI 2-1
2.2 The fundamentals and implication areas of AI 2-4
2.3 The change of maritime in AI 2-8
2.4 Future trends 2-10
References 2-13

3 Evolution of maritime autonomous surface ships 3-1

3.1 Definitions of MASS 3-1
3.2 The development of autonomy in the world 3-3
3.3 Change of MASS systems 3-8
3.4 Elements of autonomous technologies 3-12
3.5 Technical and operational constraints 3-17
3.5.1 LiDAR 3-17
3.5.2 GPS/IMU 3-18
3.5.3 Computer vision-aided systems 3-19
3.6 Future trends 3-21
References 3-23

4 Impact, development and potential for MASS 4-1

4.1 Crew careers status 4-1

4.2 Maritime industry shocks 4-4
4.2.1 The ship's characteristics design 4-4
4.2.2 System equipment 4-5
4.2.3 Construction and assembly 4-5
4.2.4 Ship's verification 4-5
4.3 The influence of port transportation 4-6
4.3.1 Wireless call 4-6
4.3.2 The pilot service 4-6
4.3.3 The work of the tugboat 4-7
4.3.4 The mooring and unmooring works 4-7
4.3.5 The cargo handling 4-7
4.4 Safety and security in MASS 4-8
4.4.1 Sensing equipment safety 4-9
4.4.2 Driving system safety 4-10
4.4.3 Control loop safety 4-10
4.4.4 Internet of Vessels safety 4-10
4.5 Legal implications of MASS 4-11
4.6 The post-COVID-19 pandemic era 4-13
4.7 Future trends 4-13
References 4-14

5 Concerns and challenges 5-1

5.1 Technological environment 5-1
5.2 Social responsibility 5-4
References 5-8

6 Conclusion and future trends 6-1

Preface

World exploration, once an impossible mission, is now common thanks to modern transportation, especially by ships. *Maritime transportation* will still be the main means of transportation in the world, according to much evidence. This trend is hardly a recent phenomenon from the last hundred years, however. Humans have been exploring Earth and using transportation by ships for thousands of years.

The maritime industry plays an important role in our modern society. Nowadays, many developed nations emphasize the importance of the development of maritime transport, and at the same time, consider it to be the key strategy in their national policy. Over the past few decades, many maritime nations have begun to see the potential of what autonomous technology can do for their national development. Some nations even consider this technology as an important task of transformation and technology upgrade. In the light of this, the author has been interested in the ocean and ships as a child. As a teenager, I have always browsed ocean-related books and paid much attention to marine and ship development. Especially when I was studying in university, I was very concerned about the development of ship technology, particularly, the integration of advanced technology and ships, which is even more amazing. However, for a long time, I have often felt that books regarding the organization of advanced ship technology are rarely seen in the market. In this sense, it may hinder many readers' interest in ship autonomous technology, not to mention setting a barrier to those who wish to further study autonomous technology for ships. As a result, this book attempts to provide readers with a systematic arrangement and introduction of autonomous ship technology, and hopes that ship technology can attract more people's attention as well.

Autonomous ship knowledge is all about ships and shipping. Another purpose for this book is to provide engineers, students, and the general public with an understanding of the field of autonomous ship technology. As readers gain complete understanding of the connotation and meaning of autonomous ships, they can thus obtain full knowledge of maritime autonomous surface ships (MASS) technology. I believe that with these conceptions, readers will be able to have clear judgments and ideas about the impact and environment resulting from these technologies.

Therefore, the author expects that after reading this book, every reader can have correct perspectives on the world situation. For example, what changes will the development of MASS bring to the world? Or, what changes will artificial intelligence (AI) bring to maritime transport? The references in this book collect relevant and important information as completely as possible. However, the study of MASS technology is quite new at this time. Since, much of the information may still be at the initial stage, the knowledge of autonomous ship technology is viewed from a bird's eye. Readers can find relevant demonstrations and references/literature in this book.

I believe those who are interested in increasing their knowledge of science and engineering will definitely be able to gain more extended knowledge from the book. On the other hand, for science or engineering readers, it is considered that this book can provide them with the direction of technological development as well as the difficulties they may encounter. Therefore, this book can cultivate readers' perspective and vision of international development.

This book has not been not completed overnight, but is compiled from accumulated years of experience and information collection. Therefore, countless people have contributed their valuable assistance and suggestions for this book. First, I am grateful to my former teachers, who gave me the foundation for professional learning. At the same time, I am also very thankful to Robert Trevelyan and Ashley Gasque at IOP Publishing for inviting me to write this book and taking the first big step to making this publication possible.

Finally, special thanks to my wife, Tina, and my two sons, Yoshi and Eason for their selfless help and devotion. My family is the inspiration of this book. Most of all, I appreciate their support and encouragement whenever I encounter difficulties in writing this book. Hopefully, after reading this book, readers can have the opportunity to make some contribution to the development of maritime industry.

Author biography

Hsin-Hung Cheng

Hsin-Hung Cheng received both B.S. degree in Marine Engineering in 2002 and M.S. degree in Mechanical and Mechatronic Engineering from National Taiwan Ocean University, in 2005, where he studied micro-electro-mechanical systems (MEMS), maritime engineering and the optics field. In 2007, he assisted the Research Laboratories of DustBuster in the Institute of Earth Sciences at Academia Sinica, in Taipei, Taiwan, which involved TOF-SIMS mass spectrometer manufacturing for astronomy research. He received his Ph.D. degree in the Department of Engineering Science and Ocean Engineering at National Taiwan University in 2013. His research focuses on biosensor applications of various spectroscopic, nano-micro-electro-mechanical systems (NEMS/MEMS) fabrication techniques including plasmonic, SERS, solar energy, and near-field optics. From 2013 to 2016, he was Assistant Professor in Department of Marine Engineering of Taipei College of Maritime Technology. In 2016, he passed the National Senior and Junior Examination in Marine Engineering. He has joined the faculty of Ministry of Transportation and Communication(MOTC), Taiwan, starting as an Associate Technician Specialist in Maritime and Port Bureau. In 2018, he has started as an Assistant Researcher and Associate Researcher in Institute of Transportation, MOTC. His work and research focuses on marine policy and management, maritime safety including marine engineering and autonomous ship engineering.

Chapter 1

Introduction

The ship's invention goes back to the very roots of civilization. Archaeologists have found simple logs of bamboo, bundles of reeds, and air-filled animal skins that prove the craft began in Asia around 3400 BC. Therefore, the appearance of a ship at a particular time and place responded to the development of the changing environment. According to the evolution of past historical processes, there are four stages of ship types, raft ships, sailing ships, steamships, and internal combustion engine (ICE) ships. The evolution of ships represents the needs of mankind for each age. In particular, under today's market trends where there is high competition, these requirements force humans to place greater demand on the development of ships.

Development of ships is more and more demanding. In the present era, we are dramatically approaching the fifth age of ships—maritime autonomous surface ships (MASS)[1].

1.1 The history of ships

The ship is one of the most common and useful transportation tools in the ocean. But do you know where it came from? If not, that's the reason why you have to read this chapter. Even though it is a common tool, people have had a hard time discovering where ships came from. Some thought its invention goes back to the very roots of civilization. In the past, people were constantly trying to create tools to help them explore the world. Among them, ships are one of the most critical tools. For those wanting to explore the world's big sights, ships make things easy.

The invention of ships has always been the most important tool influencing human transportation as well as one of the most influential tools of transportation in

[1] For readers' convenience and to avoid confusion with related nouns such as automated/autonomous/unmanned ships, and so on in the following chapters, this book uses 'maritime autonomous surface ships (MASS)' as the main definition according to IMO MSC99.

doi:10.1088/978-0-7503-3151-7ch1

the world. However, as to who invented the ship, the answer is not quite so clear-cut. In addition, researchers trying to find out the answer to this question have been studying the origins of ships. Moreover, from human remains, the early humans reveal some interesting findings. Before the wheel was invented, it is said that people were able to use materials as transportation to float on the water surface. According to the information from related ancient cultural relics, the Egyptians had invented ships as early as 3400 BC. For instance, Brooklyn Museum of New York City has in its collection a masterpiece called 'Jar with Boat Designs', that has designs of ancient Egyptians using wood to build the shape of a ship floating in water [1]. Even though this exhibit was dated around 3400 BC, people think it may be older than that.

'Jar with Boat Designs' shows ancient boats that look very different from today's as they were made from materials such as bamboo logs, and bundles of reeds. In fact, they are similar to our rafts or dugouts today. Historians have also assumed that ships certainly became popular water transportation in Egypt at that time, but the history of ships possibly goes back thousands of years, long before Egyptians established them. Even though there are different claims regarding the invention of ships, based on the above statements many facts can be verified. However, as noted above, some things should be achievable and verifiable:

1. Ships have been essential transportation in human history. After a successful presence in the human transportation scene, it is clear that ships are here to stay.
2. People depend on the convenience of ships. Ships are found almost everywhere. They have certainly come a long way to connecting the world and bringing more convenience to us.

As a result, no matter when the ship was invented, it is certainly one of the most important inventions ever made in human history, and evolution has never been stopped. Maybe we cannot imagine why the ancient Egyptians invented the ship. However, undoubtedly, it has already influenced the world by allowing people to float freely, conveniently, quickly, and safely for long distances on the water surface.

1.2 Review of the development of ships

The process of ships is a fascinating tale that spans thousands of years. Even now, it remains a fascinating tale of a person's life to be reckoned with. Generally, the history of ships developed from the earliest times to the present day consisting of four stages of ship types: raft ships, sailing ships, steamships, and ICE ships. Furthermore, MASS is now coming to the fore. Each of these ages has witnessed the history of human development. Figure 1.1 shows ships through the ages. In the following, we briefly introduce the top four most important ages of significant ship development.

1.2.1 Raft ships

In the Paleolithic age, people used rafts, bamboo logs, bundles of reeds, air-filled animal skins, and tar-covered baskets to traverse through water. These tools can help humans not only for transport but also to fish or hunt. For example, it is said

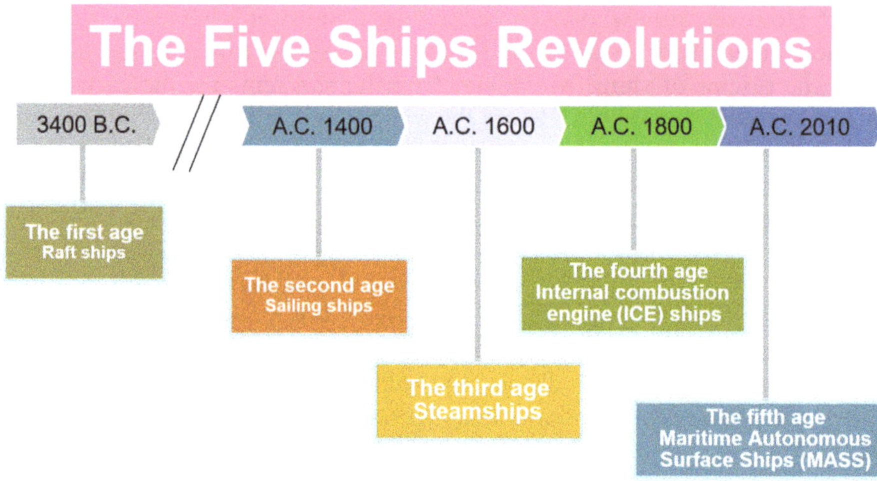

Figure 1.1. The ships through the five ages.

that the first water vehicle used by humans was called a canoe. From when humans made canoes out of huge or hard trunks, they were considered the most practical means of water vehicles. After that, when fire was gradually controlled by humans, they grilled the wood to increase the buoyancy force of the canoe. Thus, after the evolution of the canoe, there came into being the form of a raft. Different from canoes, rafts are made and bundled together in groups of trunks, branches, bamboo poles, or aloes. Furthermore, eventually the evolution of the raft meant rafts made of wood, leather, or other animal skins. At present, many people still use wood rafts or leather rafts as water vehicles in the narrow rivers of the mountains in Southeast China.

After the evolution of the canoes and rafts, people made board boats from trees and other tools by cutting and processing wood into boards. This is the beginning of humans having the ability to build a board boat and also the start of the board boat age. It was their curiosity with the ocean that aroused humans' attention on building ships. Therefore, after the launch of the board boat age, board boats entered the ocean and brought about the combination with international navigation. Board boats can still be seen in Hong Kong—for example, the Dukling.

1.2.2 Sailing ships

People have utilized wind power for centuries. According to recorded or written history, before 3000 BC, ancient Egyptians had already known how to propel a ship by wind. Vessels using wind energy as a means of ship propulsion are called sailing ships. From the view of the sailing ship invention, it is a sea-going ship that uses sails mounted on masts to harness the power of the wind. At the same time, this is also one way in which humans use Nature's power to control ships. As shown in figure 1.2, sailing ships can still be seen today.

Figure 1.2. The sailing ships of today.

The most popular age of sailing ships was between the 15th to mid-19th century. For example, in the early 15th century, Zheng-He (A.C. 1371–1433), a brilliant military leader and maritime explorer, who was a foreign diplomat in the China of the Ming dynasty sailed on seven trips to the western seas. Another example is at the end of the 15th century, when Christopher Columbus sailed his ship called Santa Maria and discovered the New World or America. And, in the mid-19th century, the latest inventions of British sailing ships and the American Clipper became the peak of sailing ship development.

During the 18th century and 19th century development of sailing ships, they expanded into more and more elements. Particularly, their features have also been gradually affected. Continuous improvement in the ship's hull, sail, mast, operation, etc, laid many important foundations for the sailing ship's development. Obviously, in Europe after the 16th century, due to the increase of displacement tonnage in sailing ships (about 500 tons), the forms of sails became more and more complicated. Moreover, when it comes to the history of sailing ships, nothing is more instructive than the sail of the Mayflower in the Atlantic in 1620, when the Puritans emmigrated to the United States from Britain. The Mayflower sail in the Atlantic is of immense significance to the subsequent history of the United States, especially as the Puritans sowed the seeds of liberty and self-government at that time, which later became the cradle of American democracy. Finally, from what we know in the 16th century, people began to use sailing ships to actively explore the unknown world. Also, they provided humans with the opportunity to exploit the sea.

1.2.3 Steamships

According to historical records, steamships replaced sailing ships in the 19th century as the preferred ships at sea. It has been found that the world's first reaction steam turbine was proposed by an ancient Greek mathematician 'Hero of Alexandria' at the beginning of the 1st century [2]. As far as we know, it was a kind of fun toy for ancient Greeks. Thereafter, thousands of years later, the first steam-powered ship was successfully introduced by an American engineer Robert Fulton in 1807. Since the first steam-powered ship was invented, the steamship's seakeeping and seaworthiness have been significantly improved, and there has been enhancement of effective operation of steamships at sea. Moreover, thanks to the advance in technology, many marine engineers saw the potential in steamships, leading to engineers involving themselves in designing, building and maintaining steamship's engines, systems and equipment. These engineering achievements allowed the steamship to obtain unprecedented and unparalleled improvements in navigation. For example, many marine engineers have been trying to navigate the open sea by utilizing steamships. Finally, in 1812, the world's first British passenger steamship successfully navigated the open sea to America, a feat accomplished by Henry Bell's 'Comet'.

From the development of sailing ships to steamships, the ship's wooden material has been replaced with metal material. This makes the ship's hull more sturdy and practical than it used to be. Similarly, more passengers and goods can be carried by the steamship's structure, making the ship's hull heavier. In addition, most steamships could already satisfy the human needs on board, and also satisfy human world exploration in the sea, as a way to connect with those living far away.

1.2.4 Internal combustion engine (ICE) ship

The ICE ship, a turning point and disruption to the traditional way of ship development, is considered the greatest global invention in this century. Nowadays, the key to making a practical ship has turned out to be ICE, which generates power by burning a flammable liquid. As early as the 1670s, Dutch scientist Christian Huygens burnt gunpowder in a cylinder to push a piston. After that, although many scientists devoted themselves to ICE application, with combustion in a cylinder not easy to control, steamships application was still of mainstream popularity at that time. But in 1860, Belgium engineer Etienne Lenoir successfully developed and produced the first ICE with natural gas based on the steam engine. Even though the efficiency was just only 2%–3%, its size could be smaller than the steamships, and so the era of steamship popularity was over.

In the past hundred years, the ICE has changed dramatically. For example, in 1862, the ICE was successfully improved by a German engineer Nicolaus Otto, to give approximately 10% thermal efficiency. After that, a German mechanical engineer Rudolf Diesel also successfully developed an engine with 26% thermal efficiency in 1893, whose theory improved combustion and improved the performance of the engine by increasing the compression ratio. Of course, this is also one of the best-known revolutionary inventions of the 20th century—'the Diesel engine'. Once Rudolf Diesel had invented the diesel engine, high-performance ICE ships also became possible. The most dramatic example was the world's first ocean-going ICE ship, MV Selandia, proposed in Denmark in 1912. As far as I know, it was built for cargo and passenger services between Denmark and Thailand. In fact, it also started a new generation in shipping that could carry more cargo over greater distances.

Today, almost 95% of shipping utilizes the ICE as the main propulsion power of ships around the world. On the other hand, although more and more means of energy propulsion power have been invented in the market, such as nuclear energy, solar energy, fuel cell, etc, it is either expensive or less efficient. What is certain is that if propulsion energy issues cannot be addressed directly, it will means that the ICE will be hard to replace quickly. Therefore, I'm absolutely sure that the ICE remains widely known as one of the greatest inventions ever to be created in the 20th century.

1.3 Future trends

Despite a long evolutionary process, ancient rafts and today's modern ships show large structural differences. Regardless, from then to now, ships have always been part of the daily life of most people around the world. Despite some debate about who invented them, many experts believe that the early ships may have originated in China or Egypt, it is people that truly made them a development of the world. These days, there is a vast range of ships: from small boat ferries, to cruise ships like traditional, holiday, and fancy, to luxury cruise ships with a special facility like an amusement park. No matter what's inside them, ships continue to be popular after thousands of years.

Since the beginning of the Industrial Revolution of the 1800s, the ship size has grown exponentially larger. For example, the world's first commercially container shipping was successfully accomplished by the 'SS Ideal X' steamship in 1956, which carried 58 containers from New Jersey to Texas. From then until today, container ships have continued to grow steadily. By 2022, the world's largest container ship in the world 'Ever Alot' owned by Evergreen Marine Corporation, was built by China State Shipbuilding Corporation (CSSC) in China. This modern marvel can carry up to 24 004 TEU (TEU is an international standard container unit) containers. Ever Alot's capacity is the equivalent of four football fields. Of course, we have already predicted that container ships may yet get bigger while the maritime industry grows more steadily in the future. Come to mention it, do you think the ship's system will become more complex while the ship size grows larger? Maybe your answer relies on your intuition, leading you to say 'of course'. To my mind, the answer is complicated, however, I believe that you will find out the answer in this book.

Although diesel-powered ships are the most common type in today's life, inventors are still hard at work developing electric-powered MASS systems. In fact, few people may know that MASS is not a wholly novel concept. The world's first unmanned ship was invented by Nikola Tesla in 1898 [3]. This ship, however, was not developed on a large scale because of technological limitations at that time. Fortunately, thanks to the rapid development of information technology (IT), MASS has been growing rapidly. Even more important, many governments in different countries had been thinking that it may also subvert the world's major transportation model. As noted above, the Mayflower is one of the most important ships in the history of the United States. However, it not only has immense significance for the subsequent history of the United States, but also brought special influence to the future of both America and Britain. Therefore, despite the immense significance of the journey, people are still expecting that the Mayflower autonomous ship (MAS) can have as great a social impact through its role in maritime transportation.

In conclusion, in this chapter we have briefly focused on the development of the ships but no detailed description was mentioned. Why didn't we give the detailed description? That's because we are going to focus on the development of autonomous ships in this book. Thus, in the next chapters, we are going to provide a comprehensive introduction of the fifth developmental age of ships—'MASS'.

References

[1] Brooklyn Museum *Jar with Boat Designs* https://brooklynmuseum.org/opencollection/objects/3276

[2] Cheng H-H 2017 *Steam Turbine* (Taiwan: Ministry of Education)

[3] Bertram V 2020 Technology trends for ships and shipping of tomorrow *Marit. Technol. Res.* **2** 1–18

Chapter 2

Background of artificial intelligence (AI)

Maritime autonomous surface ships (MASS) were not built in a day. They are the result of all scientific and technological achievements. Specifically, the key technology of MASS is credited with the perfection of artificial intelligence (AI). It is because AI has gained experience from many previous failures in the past development. Each of the failures, however, has accumulated a great deal of valuable work experience and wisdom. There is no denying that AI is still on the rise! This time, it will not only change our daily life but may also subvert the world's major maritime transportation model.

2.1 Review of AI

Do you remember the movie 'The Terminator'? What comes to mind when someone brings up the terminator robots? Many movie fans think of negative things, such as robots used in this movie for bombings and surveillance, or personal robots causing danger to planes when flying high in the sky. On a positive note, ask any employer what kind of staff they want, and they will probably say efficient workers who never take time off. While this is understandable, humans cannot always meet these standards. Human limitations have therefore led to machines replacing them in workplaces for well over a hundred years now.

The above two examples are about robots in human imagination. In order to realize these human imaginings, the robots need to be changed significantly. For example, to make robots work in more diverse ways, and be faster, more regular, and more efficient than humans, these robots themselves need more complicated hardware and software facilities. Therefore, as AI technology becomes increasingly capable, the systems of AI have gradually become smarter and more intelligent [1–4]. In terms of ships, many studies have been discussing the application of AI in ships [5–8], and also proposed some concepts of the possibility of MASS or autonomous sailing of ships.

doi:10.1088/978-0-7503-3151-7ch2

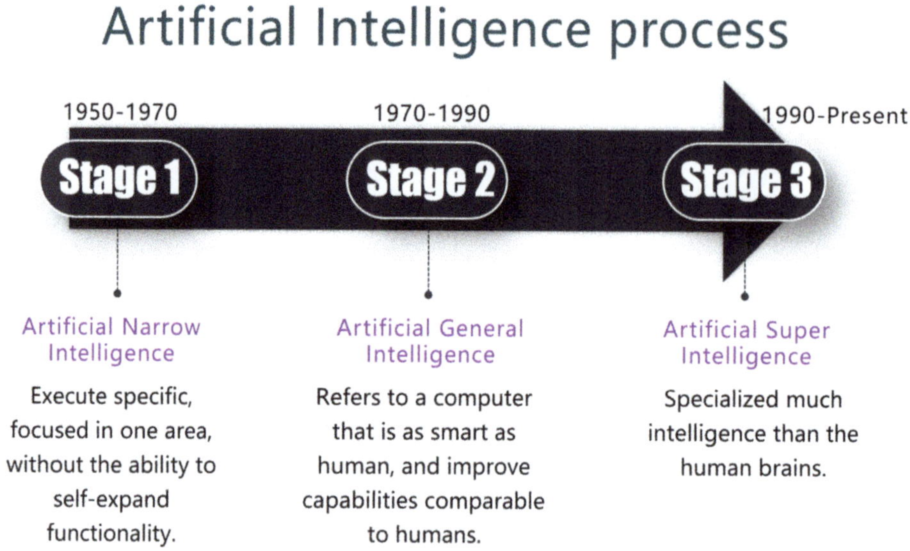

Figure 2.1. The artificial intelligence process.

First of all, many people may think that AI must be new innovations and revolutionary projects, but actually that is not the case. In fact, the development of AI is not as smooth as it seems. Before AI, electrical and automation works developed over 30–40 years and have been growing rapidly. Thus, the mature development of automation can be seen in many areas of life, especially in industrial manufacturing.

The programming language was input by humans into the computer, allowing the computer to drive the motor driver of the mechanism based on the programming language. Eventually, it completes the action set by human beings. For AI, different from the automated mode operation, the performance of AI implants the wise knowledge of humans into the robot's programs (the robot's control systems). Finally, it completes the things that humans want to accomplish, predicting the results that humans want to obtain, and even solves the problems that humans want to solve [9, 10]. Have you ever stopped to think about the history of the AI process? Probably not, you just want to enjoy the benefits it brings. When you've finished enjoying it, check out the following process.

The AI development process is full of ups and downs. Figure 2.1 shows the AI development process. Generally speaking, the development of AI can be divided into three: artificial narrow intelligence (ANI), artificial general intelligence (AGI), and artificial superintelligence (ASI) [11]. The first AI process was ANI. As early as the 1950s in the US, the computer had only just started. Gradually, the computer has greatly increased the usability of human work. Nowadays, numerous applications in the computer field, including, information technology, business and finance, health science, architecture and construction, etc are becoming more popular across the world. There is no denying that the computer is a useful tool to realize human dreams and create infinite possibilities to help people handle difficult, cumbersome, anxious, desirable, dangerous, dull, and depressing problems efficiently.

However, when really expecting that computers can think like humans to solve problems that they have encountered, we must refer to the concept of 'AI' proposed by McCarthy *et al* [12] at the Dartmouth conference in the United States in 1956. At that time, this concept was still a vague imagination for humans, and they could not totally capture the unique essence of AI. Humans expected to create a robot or computer with intelligence like a human. Besides, from the period of the 1950s to the 1970s in AI development, scientists tried to put concepts from symbolic logic into computers to make decisions. For example, the most typical example of the category is Boolean algebra [13]. However, at that time, people expected to use symbols to achieve the behavior of computer thinking, and even purely believed that computers with logical reasoning could realize AI. After years of hard work, humans finally found that putting symbolic logic into a computer is still far from reaching the level of AI. Therefore, the first periods of AI technology development began to sink for a while.

AGI [14] can be traced back to the period from the 1970s to 1990s, when AI experienced the rapid growth of information technology and the development of ANI. The development of AI has scientists indulging in new AI processes, including human knowledge and experience. The form of human behavior is also highly valued, such as experience, life and habits, etc. ANI successfully portrays the spirit of this age by characterizing scientists' pursuit of these pleasures.

During the AGI stage, scientists also tried to integrate human knowledge into computers. Specifically, at that time, the experience and knowledge of experts in many domains were converted to all computing languages, and enabled computers to understand and derive meaning from human experience and knowledge. Of course, scientists still hope that the experience and knowledge of experts can be human-like, transforming masses of creative and constructive ideas, earning the well-deserved name of 'artificial intelligence'. In fact, to think is easy, however, it is extremely difficult to realize it. For example, if we want to develop a medical diagnosis system with AI functions to determine and explain what disease or condition a person's symptoms are caused by, first of all it is necessary to construct the medical diagnosis database which collects the experience and knowledge of many doctors, and then, like a doctor, determine the treatment and dosage of the prescription. Finally, let the machine learn by itself.

In fact, this development not only consumes a lot of time, but also human resources, facilities, finances, etc. Moreover, as humans come to understand the implications of integrating human knowledge, they eventually see it as an ordinary automated process. Thus, the technology in the AGI stage still cannot simulate human thinking to make adequate judgments, and the temporary setback in the AGI period still frustrates scientists.

The third period of AI is called the 'machine learning period (ASI) [15]', which developed approximately from 2010 to the present. In this period, humans hope to put whatever they see, hear and speak all into computer programs. They create, compile, and organize a great amount of professional historical knowledge, experience, and data from experts in various fields, so that the computer can self-summarize and learn relevant rules. Especially those specializing in deep learning, humans hope to utilize these big data analysis sets for machines, and then the machines can learn knowledge by themselves in automatic learning mechanisms. Fortunately, thanks to the significant improvements in computer hardware technology, the big data analytics skills have become more and more usable over time, as well as being realistically accessible in some form for use by companies. For example, as Alpha Go defeated the world Go king, Lee Sedol in 2017, it became a specific form of deep learning. In brief, the big data analytics sets or deep learning can almost help machines improve AI by the process of collecting, examining, and analyzing large amounts of data, and other related tasks. Why not follow their development? Take what you've learned about deep learning and dispense some knowledge to maritime fields.

In sum, AI has been developing for almost 7 decades, it is still progressing continuously on the basis of our hopes. In any case, AI may quickly break through with impact, as what was once shocking can quickly become mundane. Hence, we absolutely believe that one day AI will bring more amazing worlds to humans because of its success.

2.2 The fundamentals and implication areas of AI

The world is always changing. New technologies are popping up all the time, and AI is constantly evolving. Now it is time to give the sense of AI a makeover. Indeed, one of AI's most iconic applications should widely be applied in various fields of today's society, rather than choosing which fields to be used in. This sense emphasizes the importance of the application of AI. For example, ever since computers were invented, initially, people were thinking what can computers help us do? Where can computers be used? How can computers help people quickly solve real-world problems and improve efficiency at the same time?

Obviously, after more than 70 years of evolution, computers have played an important role in various fields, such as solving cumbersome, repetitive, tedious, frustrating, and mass calculation problems for people. Moreover, AI is the same now as computers were when they emerged. It may also assist people to address similar problems in various fields in the future, even accelerate people to address more difficult problems. But how does AI accomplish this? Next, let us take a look at the learning architecture of AI to see this amazing technique once again.

Good learning is half the battle. This is especially the case for learning problems. Humans have always played a key role in carrying out relevant behaviors in traditional diagnosis and decision-making processes. In other words, once human information is transmitted in the brain or neuron activity, the human brain will process information and allocate attention according to experience, knowledge and then make decisions as accurately and quickly as possible. Apparently, the learning evaluation procedures have been preliminarily understood. Figure 2.2 shows the difference between machine learning with feature extraction and deep learning [16]. In machine learning, when combining the expert experience and traditional classification,

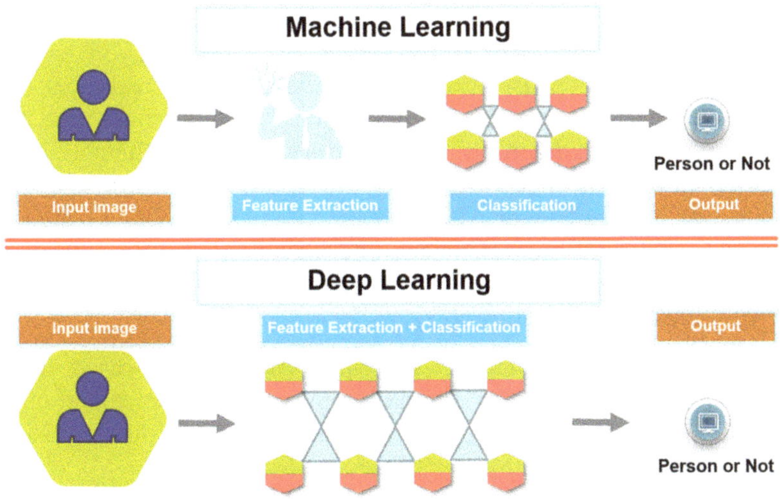

Figure 2.2. Machine learning and deep learning system.

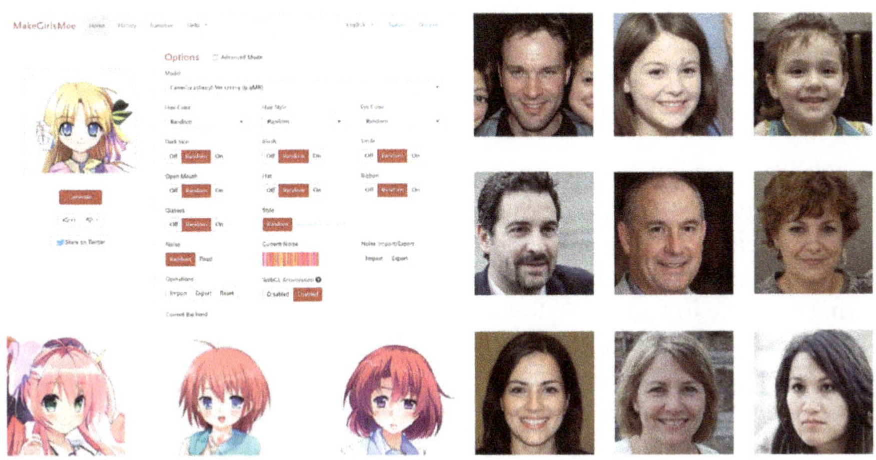

Figure 2.3. The AI websites: (a) MakeGirlmoe http://make.girls.moe/#/; (b) thispersondoesnotexist http://thispersondoesnotexist.com/.

classification can be regarded as a neural network-like model to assist humans in diagnosis. Further, machine learning also helps to accelerate the traditional human brain process for operating the diagnosis. The neural network model mentioned here is to calculate the data layer by layer, and then eventually obtain the expected result. For deep learning, when the information is input, the diagnosis and decision-making results can be obtained through repeated feature extraction and classification.

The above example provides a tough situation indicating that AI has pushed all human performance into the background. However, deep learning can still be frequently seen in our lives. Figure 2.3 shows pictures generated by AI. Take the MakeGirlmoe website as an example, we can feature various traits of humans on the website, such as: hair color, hair style, eye color, or dark skin. Through the function of this site, we can get a picture of a cute cartoon girl. Of course, you may also try to surf the 'thispersondoesnotexist' website to see AI-generated real faces. Wow! It's true! What you see is not a real human, all the faces in the website were created by AI. That sounds horrible, doesn't it? Moreover, if you find anything wrong or weird about the human face, or there are strange patterns on other side of the faces, don't doubt it, it is just a deep flaw of AI.

Despite AI's incredible development and meaning, there's more to it. For example, the following three definitions are all about an AI project, but they express different meaning and applications [17]:

1. AI, a general term definition. Applying various techniques in the software and hardware of machines, simulating human intelligence and understanding. This includes machine learning, deep learning, computer vision, visual neuroscience, natural language processing tools, etc.
2. Machine learning (ML) is a part of AI. It applies to a variety of mathematical (such as probability, calculus, linear algebra, etc) and statistical methods to let machines learn by themselves and train themselves for supervised learning, rather than following common procedural rules.
3. Deep learning (DL) is a deeper part of ML, which is also an evolution of ML. The only difference from ML is that DL may obtain the results of predictions and judgments beyond human intelligence.

So far, we can see that AI is only a rough general term including ML and DL, a meaning that does not have a standard descriptive form. However, what are the different applications between the meaning of ML and DL in AI? In order to realize that AI applications are full of springing surprises in everyday life, please see the following example in figure 2.4. Read on to learn several surprising facts about AI.

First of all, a complete, comprehensive database system of artificial neural network is necessary to develop AI. These database systems store essential big data about information sets. As can be seen in figure 2.4, the neural network process in ML can be read from top to bottom. Here, if a ship image was observed by the camera, the input of the ship image must be determined in the input layer of ML. Next, in the second internal layer (called hidden layer), when we don't know the input feature of the ship is a variable, we must pre-provide the feature to distinguish the ship. And then, through the different weights applied to obtain the feature variable of the ship, we obtain the results of humans' expectations. In this layer, ML must do some complicated math calculation, learn by itself and continuously complete the training. Last, In the output layer of the third layer that contains all the possible outputs, it begins to compare the differentiated ratio of ships and other transportation (such as, vehicles, airplanes, scooters, etc) in ML. In fact, ML cannot easily distinguish images of ships from other people, even though the results of weighted calculations are not that clear. In addition, sometimes, it may still be impossible to approach the expected results for humans after ML. Therefore, DL is to address this problem.

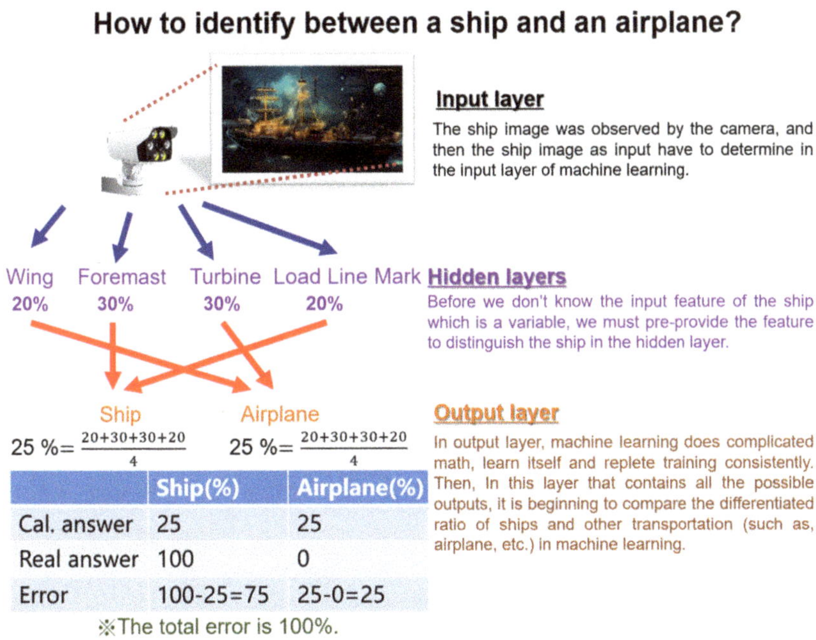

Figure 2.4. Machine learning: common learning architectures.

As can be known from the above, in order to make ML continue to develop well, the database must contain complete, serviceable, comprehensive, and characteristic data. Specifically, these database systems must be summarized, sorted, and classified as much as possible, so that under ML condition, the more accurate it is the better the performance. At this time, as a part of ML that due to deep learning is the features and weights, can be further modified in all hidden layers based on ML. Therefore, after an interactive multi-task training process in both ML and DL, we can gradually obtain the results processed by AI. Furthermore, the result of the ship image that is judged by AI may also be similar to humans' expectations. For example, figure 2.5 shows an example of DL.

In this example, many subjects were observed by the camera (such as, vehicles, airplanes, ships, etc), but only the ship was chosen as this input layer of DL. After that, in the hidden layers, we must first provide features of ships, vehicles or airplanes to obtain the results of humans' expectations. Likewise, based on the results of ML, DL also must do some complicated math, as well as consistently adjusting the features and weights. And then, DL compares the ship with each means of transportation to obtain the feature of the differentiated ratio of the ship. It is worth mentioning that the percentage error of the ship dropped from 100% to 60% after calculating with ML and DL. Of course, whether it is ML or DL, the more the database systems with better calculations, the better the performance will be.

Although AI experienced many setbacks and challenges, humans have never let these setbacks and challenges stop its development. To humans, fearlessness is developed through hard work, in spite of the things that may make humans fail.

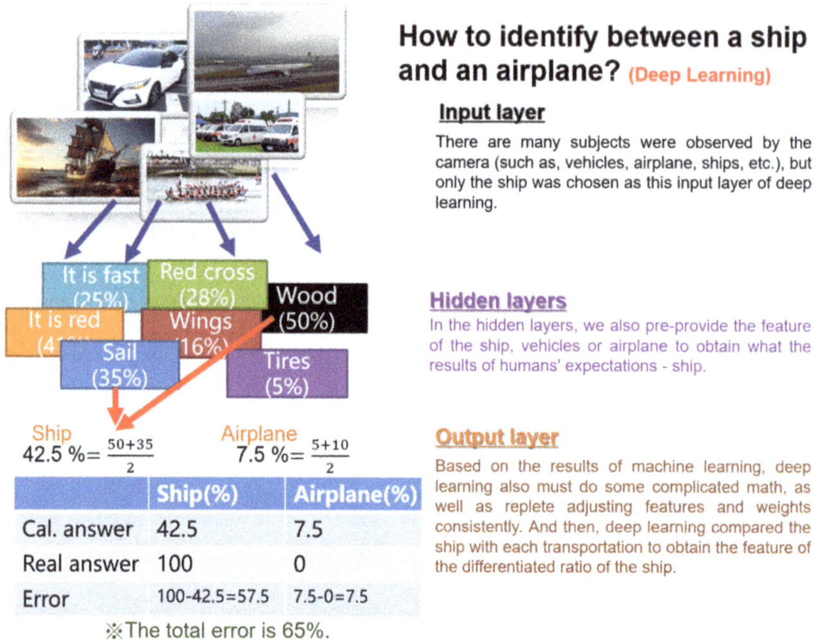

Figure 2.5. Deep learning: common learning architectures.

Clearly, no matter when the development of AI is mature enough, what we can expect is that whether AI will be able to get closer to human intelligence depends on human desire and effort.

2.3 The change of maritime in AI

Although AI is still developing, its potential is promising. Many individual aspects of AI have interesting applications behind them, one of which is self-driving or autonomous driving. Take maritime transportation, for example. Every day, there are many ships sailing at sea. It would be so exciting if the ship could be handled by AI in the process of entering and leaving the port. So, how does AI play a key role in the transportation of shipping process? Let's take a look at what new possibilities and opportunities these ships with AI applied can bring.

Generally speaking, as shown in figure 2.6, there are four stages during the operation of MASS ships 'arrival at the port' and 'departure from the port'. First of all, 'pre-navigation preparation' is the first stage. In this stage, the ship's crew has to pre-plan the navigation schedule based on the navigation database, voyage planning, and discuss all the matters that would be required for other related work. Such as, to ensure safe maneuvering, cargo operation and safe berthing of ships at the port. Apart from the above-mentioned procedures, all important navigation equipment etc, must be tested before the actual maneuvering process begins. The second stage is 'navigation monitoring'. In order to grasp the working environment of the ship while sailing, which also includes the procedures of various sensing and

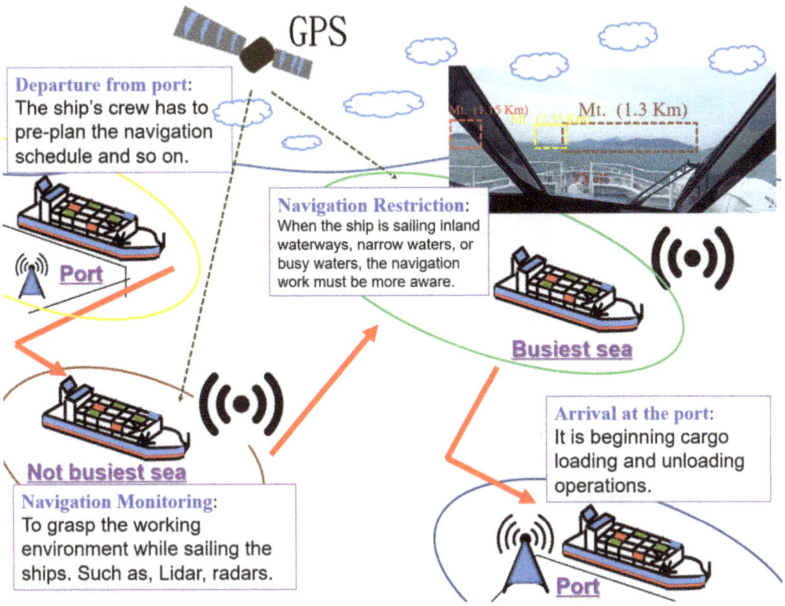

Figure 2.6. Application of artificial intelligence in the maritime environment.

monitoring equipment mentioned earlier, for example, infrared cameras, radars, or related sensing components etc must be tried before the actual process begins. At this time, it must rely on satellite communications to keep in touch with the shore [18]. The third stage is 'navigation restriction'. While the ship is sailing in inland waterways, narrow waters, or busy waters, navigation work must be more aware at this time. Therefore, whether it is navigation environment identification, sensor sensitivity, or other navigation information, signal and information processing must all be significantly faster and safer. Examples of this would be when navigating on the Panama Canal, or the Suez Canal.

The last stage is called 'ships arrival and departure from the port operations'. Before the ship approaches the port, preparations must be made for the work to be carried out in sequence. Once the ship completes berthing in the port working area, cargo loading and unloading operations begin. Finally, it is also ready to prepare for the next sailing.

As can be seen from the above four stages, when we compare MASS with traditional ships, it is found that there's not much of a difference between the traveling process of MASS and traditional ships. Technologically, the only difference is that MASS must need more components of real-time sensing, communication, monitoring systems etc than traditional ships. This way MASS traveling might (or must) be completed with AI algorithms [19–21]. For example, Wright [6] discussed reducing sea accidents by using AI as well as completing all the travel. Indeed, these tasks seem like a simple process; however, the work required may be beyond our imagination. In addition, we can also understand that when the components of real-time sensing, communication, monitoring systems etc of

(a) Road Transport Route (b) Sea Transport Route

Figure 2.7. Transport modal shift (road to sea).

MASS replace the human sensory organs, AI will soon be able to also replace human judgment. Undoubtedly, this is indeed a big challenge.

Although it is still a great challenge to replace human-controlled ships with AI, there are still many benefits. For example, Munim [22] has evaluated some related benefits when realizing the autonomous ship. Such as, an autonomous container ship can save US$357 per day on fuel cost; for a full battery-driven autonomous container ship, there are no carbon dioxide and nitrogen oxide emissions. In addition, it can also bring significant benefits to reduce the number of casualties of the crew and avoid the threat of pirates. Fortunately, to realize these significant benefits, Norway has already translated MASS vision into a series of specific actions that we can begin on as soon as possible. Norwegian shipbuilder Vard has handed over the world's first battery-powered autonomous container ship—Yara Birkeland to Yara International. In addition, figure 2.7 shows the transport mode shift of YARA Brikeland process (a) road transport route and (b) short-sea-shipping route, between the same two ends Prosgrunn and Larvik. In the past, when YARA International transported fertilizer from Porsgrunn to both Brevik and Larvik, it required more than about 40 000 truck journeys to these two populated urban areas every day. At present, with the realization of the autonomous ship in 2021, land transportation can be replaced by waterway transportation, thereby reducing noise, dust, as well as emissions of carbon dioxide and nitrogen oxide.

As can be seen from the above, although currently developing MASS and AI may be a little challengeable, it really brings a significant increase to the efficiency of maritime industry. Therefore, with AI allowing MASS to take away travel risks for humans and assisting them in doing repetitive jobs, human beings are one step closer to realizing the dream of MASS travel. In the near future, seeing MASS travel in the sea may not be out of reach for us.

2.4 Future trends

There are many ways in which AI can be developed while at the same time, it is important to understand its many limitations. First of all, up to now, everything has not been as successful as people thought it would be. Through some development process, it was found that integrating human experience and knowledge can help AI develop useful skills. Crews who work in maritime fields of AI learn how to react to various climate and hydrology situations, for example. Therefore, it is a valuable and significant skill when AI faces challenges in any situation. What's more, many researchers think that developing AI with human experience needs more creativity and friendliness. This is especially the case when some people may argue that AI is not progressing very well, as many of them assume that AI requires to learn more about human emotions, attitudes and social abilities. In fact, inputting human emotions, attitudes and social abilities into AI is not an easy task. Therefore, for the development of AI, every adversity, every failure and every challenge is accompanied by innovative thinking, which brings greater benefits. Moreover, despite the fact that AI is at an immature stage now, it has given us some things worth thinking about. What are they? Let me give you some examples to illustrate:

1. The way humans think.
   AI-powered humanistic thinking is nothing new. The standard method is to accumulate a great deal of valuable work experience and wisdom, convert it into programming language, and convert it back to complete the actions set by human beings. However, this method has obvious limitations when we notice humans trying to use mathematical methods to describe the process of human behavior development. To address this issue, many researchers are working on some new open source systems that can allow human beings to instantly change from one programming language to another.

   So far, AI-powered humanistic thinking systems are only able to provide rough decisions between work experience and programming language, with no standard written form of working experience. It integrates several approaches to perform this task, including converting mental modes into action inputs and then generating them, as well as training AI mode by using the programming language as an intermediary between work experience and programming language. Currently, the system can translate only one action at a time. According to many researches, however, this is just the first step toward achieving instant switching between human thoughts. Simply speaking, importing human experience and knowledge into computer programs may take a long time before achieving human decision-making.

   Many researchers hope its system can be used to facilitate communication in both the real world and the virtual worlds. By breaking down programming language barriers, such AI technology will hopefully lead to greater understanding and collaboration among speakers of different programming languages.
2. Integration of various fields.
   Imagine sailing on the high seas on the newest AI-powered MASS voyage where you can scan your face or fingerprint to unlock the ship's propeller. Of course, besides developing AI technologies, integration of a database in various fields is second to none. Creating a database is a huge project which organizes the collection of data and structured information. In addition to

acquiring the ability to collect and store a capacity of information, the database must also have comprehensive expert knowledge and experience. In fact, in the process of AI development, creating a database like this is not that easy to address, not to mention the knowledge and experience that have been lost or are about to be lost. For example, in the maritime field, many advanced maritime nations have created a marine accident database to figure out the complicated reasons for maritime accidents and maritime casualties [8, 23]. And the collected data information will determine the processing and decision-making from experts' experience. However, it seems that we still have a long way to go when it comes to collecting, creating, and storing the capacity of information to reduce maritime accidents or maritime casualties by AI.

3. Human acceptance.
   Have you ever seen the movie 'Blade Runner'? How much could you accept from a robot? For example, when you go to the movie with your family, could you accept the other robots lining up with you to buy tickets? Or when you drive to work, could you accept that there are many unmanned self-driving cars with you on the highway? If you can accept the above, how about the idea of an unmanned self-drive plane to take you abroad? Maybe you already have the answer in your mind right now. Judging from the current situation, human beings are not willing to accept such things to some extent.

   On the positive side, thanks to the invention of AI, people can see what our maritime application will look like from every angle. Although AI cannot replace human thinking completely, AI can assist humans to check autonomous control systems and detect early signs of operation, such as programming errors. On the other hand, AI can help people manage and arrange travel or navigation records so that all the information can be stored in a cloud database.

   AI has many other applications concerning the maritime industry. Although the realization of MASS may fill you with excitement, there are still some issues to be overcome before the technology can be completely embraced. The foremost concern is making sure these MASS are safe, particularly when sailing in narrow waters or busy waterways, which is the most dangerous part of any voyage. Without strict regulations, you could not see MASS whizzing around the sea. Another issue is who will actually be driving these MASS? According to related studies [24], MASS operation mode will be divided into two types: ship-side and shore-side. However, the operators of the shore control center would require thorough training if safety standards are to be met. Then again, it seems that many of the companies developing MASS are interested in making autonomous sailing, meaning that this concern may soon be solved.

   In short, AI is an incredible technological advancement. There is no denying that it will be utilized in our future world soon. For example, since the invention of ships, human error has still concerned ship manufacturers and ship owners [25, 26]. Many people are looking forward to using MASS functions, technologies, etc to replace human error in ships, making shipping safer and more efficient. The good news is that autonomous/unmanned ship technology is looking to change these situations through the technological wonder of ML, etc.

References

[1] Kumar D and Meeden L 1998 A robot laboratory for teaching artificial intelligence *ACM SIGCSE Bull.* **30** 341–4

[2] Hamet P and Tremblay J 2017 Artificial intelligence in medicine *Metabolism* **69** S36–40

[3] Liu R, Yang B, Zio E and Chen X 2018 Artificial intelligence for fault diagnosis of rotating machinery: a review *Mech. Syst. Sig. Process.* **108** 33–47

[4] Ogle A and Lamb D 2019 The role of robots, artificial intelligence, and service automation in events *Robots, Artificial Intelligence, and Service Automation in Travel, Tourism and Hospitality* (Emerald Publishing Limited) pp 255–69

[5] Zhang T, Li Q, Zhang C S, Liang H W, Li P, Wang T M and Wu C 2017 Current trends in the development of intelligent unmanned autonomous systems *Front. Inf. Technol. Electron. Eng.* **18** 68–85

[6] Wright R G 2019 Intelligent autonomous ship navigation using multi-sensor modalities *TransNav: Int. J. Mar. Navig. Safety Sea Transp.* **13** 503–10

[7] González-Cancelas N, Serrano B M and Soler-Flores F 2019 Seaport sustainable: use of artificial intelligence to evaluate liquid natural gas utilization in short sea shipping *Transp. J.* **58** 197–221

[8] Munim Z H, Dushenko M, Jimenez V J, Shakil M H and Imset M 2020 Big data and artificial intelligence in the maritime industry: a bibliometric review and future research directions *Marit. Policy Manag.* **47** 577–97

[9] Nilsson N J 2014 *Principles of Artificial Intelligence* (Morgan Kaufmann)

[10] Ghahramani Z 2015 Probabilistic machine learning and artificial intelligence *Nature* **521** 452–9

[11] Bengel D 2020 *Organizational Acceptance of Artificial Intelligence* (Springer Fachmedien Wiesbaden)

[12] McCarthy J, Minsky M L, Rochester N and Shannon C E 2006 A proposal for the dartmouth summer research project on artificial intelligence, August 31, 1955 *AI Mag.* **27** 12–2

[13] Muller D E 1954 Application of Boolean algebra to switching circuit design and to error detection *Trans. IRE Prof. Group Electron. Comput.* **EC-3** 6–12

[14] Goertzel B and Pennachin C 2007 *Artificial General Intelligence* (Berlin Springer)

[15] Batin M, Turchin A, Sergey M, Zhila A and Denkenberger D 2017 Artificial intelligence in life extension: from deep learning to superintelligence *Informatica* **41** https://research.ibm.com/publications/artificial-intelligence-in-life-extension-from-deep-learning-to-superintelligence

[16] Krishna S T and Kalluri H K 2019 Deep learning and transfer learning approaches for image classification *Int. J. Recent Technol. Eng.* **7** 427–32

[17] Bini S A 2018 Artificial intelligence, machine learning, deep learning, and cognitive computing: what do these terms mean and how will they impact health care? *J. Arthroplasty* **33** 2358–61

[18] Atayero A A, Alatishe A S, Matthew L K and Orya M K 2011 Satellite communications: impact on developing economies *J. Emerg. Trends Comput. Inf. Sci.* **2** 574–9

[19] Montealegre S, Hammer K and Dissosway M 2005 *U.S. Patent Application No.* 10/703, 290
[20] MacKinnon S N, Weber R, Olindersson F and Lundh M 2020 Artificial intelligence in maritime navigation: a human factors perspective *Advances in Human Aspects of Transportation: Proc. of the AHFE 2020 Virtual Conf. on Human Aspects of Transportation (July 16–20, 2020)* (Springer International Publishing) 429–35
[21] Zhang C, Guo C and Guo M Z 2020 Information fusion based on artificial intelligence method for SINS/GPS integrated navigation of marine vessel *J. Electr. Eng. Technol.* **15** 1345–56
[22] Munim Z H 2019 Autonomous ships: a review, innovative applications and future maritime business models *Supply Chain Forum: Int. J.* **20** 266–79
[23] Mirović M, Miličević M and Obradović I 2018 Big data in the maritime industry *Naše More* **65** 56–62
[24] Im I, Shin D and Jeong J 2018 Components for smart autonomous ship architecture based on intelligent information technology *Procedia Comput. Sci.* **134** 91–8
[25] Oluseye O O and Ogunseye O O 2016 Human factors as determinants of marine accidents in maritime companies in Nigeria *J. Marit. Res.* **18** 1–11
[26] Galieriková A 2019 The human factor and maritime safety *Transp. Res. Procedia* **40** 1319–26

Chapter 3

Evolution of maritime autonomous surface ships

Looking back at the origins of maritime autonomous surface ships (MASS), it is shocking to realize that they have only been about for a few decades. During the past decade, breakthrough progress has been made in the development of autonomous technology. For example, significant changes have taken place in crewless ships, changes in ship control, autonomous navigation route planning, and the port and ship operation modes. These have currently become more focused with the development of MASS. As the development of MASS is flourishing all over the world, many governments are thinking that it is a significant task. It can be imagined that the application of MASS will once more subvert the concept of using traditional ships in the future [1]. It goes without saying that the realization of MASS is a great way to support people in improving maritime transportation efficiency.

3.1 Definitions of MASS

Many people believe that the International Maritime Organization (IMO) has only noticed the development of automated ships in recent years, but that is actually not the case. The Inter-Governmental Maritime Consultative Organization (IMCO), before IMO changed its name in 1982, had discussed the title 'Automation in Ships' as early as 1964 in a report and document on automated ships (IMCO Doc MSC VIII/11). However, with the rapid development of semiconductor and information technology (IT) in recent years, automated ships have gradually gained attention. For example, the European Union, the United States, the United Kingdom (UK), and Asian nations have all begun to develop technologies related to automated ships, and regard automated ships technology as the focus of future maritime development. Because of this, in the past decade, the development of automated ships that have autonomous/unmanned ships technology have gained more attention and new opportunity.

A brief description of the work done by IMO to develop automated ships in recent years is as follows. Figure 3.1 shows MASS development status of international maritime organization.

doi:10.1088/978-0-7503-3151-7ch3

Figure 3.1. MASS development status of international maritime organization.

1. June 2017, the 98th session of IMO's Maritime Safety Committee (MSC98), a wide range of topics was on the agenda, including the UK, Denmark and other maritime nations of Europe putting forward the studies topic of ship automation, and putting forward MASS at the meeting, and the consideration of a regulatory scoping exercise (RSE).
2. May 2018, the 99th session of IMO's MSC (MSC99) endorsed a framework for an RSE, including MASS being defined as 'as a ship which, to a varying degree, can operate independently of human interaction, preliminary definitions degrees of autonomy, as well as a methodology for conducting the scoping exercise and a plan of work'.
3. December 2018, the 100th session of IMO's MSC (MSC100) approved the framework and methodology for the RSE on MASS, and identified the four degrees of autonomy. The degrees of autonomy identified for the purpose of RSE are:

    (1) Degree one: Ship with automated processes and decision support: seafarers are on board to operate and control shipboard systems and functions. Some operations may be automated and at times be unsupervised but with seafarers on board ready to take control.
    (2) Degree two: Remotely controlled ship with seafarers on board: the ship is controlled and operated from another location. Seafarers are available on board to take control and to operate the shipboard systems and functions.
    (3) Degree three: Remotely controlled ship without seafarers on board: the ship is controlled and operated from another location. There are no seafarers on board.
    (4) Degree four: Fully autonomous ship: the operating system of the ship is able to make decisions and determine actions by itself.

4. June 2019, the 101th session of IMO's MSC (MSC101) approved Interim guidelines for MASS Trials (MSC.1/Circ.1604). The guideline also indicates that MASS risks associated with the trials should be appropriately identified and measures to reduce the risks, to as low as reasonably practicable and acceptable, should be put in place.
5. May 2021, the 103th session of IMO's MSC (MSC103) completed an RSE for safety treaties. This was designed to assess existing IMO instruments to see how they might apply to ships with varying degrees of automation.
6. April 2022, the 105th session of IMO's MSC (MSC105) commenced the development of a non-mandatory MASS Code, with a view to adoption in the second half of 2024 as the first stage. In addition, based on the experienced gained in the application of the non-mandatory MASS Code, a mandatory MASS Code will be envisaged to come into force in 2028.
7. October 2022, the 106th session of IMO's MSC (MSC106) commenced the development of a goal-based instrument for MASS.

Judging from IMO's work on MASS, many important foundations have been laid for its development, such as definitions of MASS, degrees of autonomy, as well as the completion of an RSE. In fact, these works are of great significance to MASS. For example, before determining the definition of MASS, many nations often referred to similar terms like unmanned ships, intelligent ships, smart ships, self-driving ships. These nouns were not unified until IMO defined MASS. Thanks to IMO, when it comes to MASS, we are thinking the same.

3.2 The development of autonomy in the world

Reviewing from the level of activity in developing MASS in the past, a number of organizations have also been working to give ordinary people the opportunity to navigate on the high seas in unmanned, autonomous boats. Among them, the development of unmanned ships in Europe is more active. These organizations have grown rapidly in the European Union. Next, let us introduce the important works and organizations in autonomous technology.

1. Maritime Unmanned Navigation through Intelligence in Networks (MUNIN)
   The earliest part of the world to notice the development of autonomous ship was Europe. In 2007, the Seventh Framework Programme of the European Commissions had already discussed issues related to unmanned ships. In 2013, it launched the Maritime Unmanned Navigation through Intelligence in Networks (MUNIN), whose aim is to develop and verify a concept of an autonomous ship. In 2015, MUNIN released the 'Qualitative Assessment' document, which including: Identification of main risks, flag states view on autonomous ships, and external stakeholder consultation. Especially, it was the earliest document to report the results of an in-depth qualitative and

quantitative analysis of the concept of autonomous ships, which covers safety and security impacts, economic impacts and application areas of law [2].

This project took three years and ended in 2015. According to the result of the final document—'Research in maritime autonomous systems project. Results and technology potentials', the EU has already discussed the future development of the maritime industry regarding competitiveness and innovation, while also considering the operation of an unmanned merchant ship and assessing its technical, economic and legal feasibility. Obviously, these efforts have served as a fundamental sample for feasibility studies and assessments of maritime autonomous systems development.

2. The UK in Advanced Autonomous Waterborne Applications (AAWA)
The UK is one of the nations around the world that is vigorously developing MASS functions. According to the campaign of MASS development, several government departments and enterprises in the UK are jointly developing MASS functions with a goal to take the lead in the maritime area. Among them, the AAWA launched by Rolls-Royce was formed through collaboration with the universities, ship designers, equipment manufacturers and classification societies, etc. It is committed to solving the problems that are hindering the development of autonomous ships from the point of view of economy, society, law, supervision and technique as well as providing reviews, making policies, and making preliminary ship designs for the next generation of autonomous technologies.

In 2016, Rolls-Royce published a white paper named 'Remote and Autonomous Ships—The next steps'. The white paper explores the development of autonomous ships in the long term from technology, law, safety and security, and business perspectives [3]. Moreover, it also proposes many important development plans on the technologies for maritime situational awareness and navigation, legal implications of autonomous shipping, safety and security, and innovations markets, etc, whose important outcomes are also supported by many fields and stakeholders. For example, in the technologies for Maritime Situational Awareness Project, Stella, a 65-meter-long double-headed ferry, operates from the ports of Korpo and Houtskär in the Archipelago Sea on the southwestern coast of Finland. For Stella, it conducts a series of sensor arrays experimental research and engineering applications for the new autonomous system. Finland's ESL Shipping Ltd has brought together the AAWA initiative, which can provide assistance to explore the influences of autonomous ships in the short-distance cargo field. This is a concrete achievement. Therefore, it can be seen that the white paper has been a valuable reference for future real ship tests.

3. Norway
Norway, like the UK, is one of the more active nations in the development of autonomous ships. In October 2016, The Norwegian Maritime Administration (NMA), The Norwegian Coastal Administration (NCA), The Federation of Norwegian Industries and MARINTEK (now SINTEF

Ocean) established the forum 'The Norwegian Forum for Autonomous Ships (NFAS)' in Oslo. The objectives of the forum include: (1) providing a platform to strengthen the cooperation between members, researchers, authorities and others that are interested in autonomous ships and their use; (2) contributing to the development of common Norwegian strategy for development and use of autonomous ships, which contains research and innovative strategies, principles of regulation and approval, as well as national testing facilities; (3) becoming a common discourse leader in the development and use of autonomous ships based on the goal of interest-based organizations; and (4) strengthening the international connection and influence of Norwegian interest groups within the area of autonomous ships.

As mentioned previously, established by NFAS, the forum is offering promising opportunities for international cooperation. It is worth mentioning that in terms of the importance of the above development, NFAS is also aware that at present, no clear guidelines and requirements are provided for the risk and hazard potential of autonomous ships. This also broadly implies that we should invest more in safety works. Among them, advanced ICT (Info-Comm Technology) systems including new sensors (RADAR, LiDAR, and visual cameras, etc) may be the key factor to raise the costs. Another aspect is the regulatory scoping supervision issue, since the current framework of maritime regulatory was developed before MASS was established. Thus, the wide range risk definition, hazard issues, and regulatory gap need to be integrated into all existing international and national maritime regulatory frameworks. Many Norwegian enterprises show outstanding performance. A brief explanation is as follows:

(1) Kongsberg Maritime
    Norway Kongsberg Gruppen is a world leader in advanced technology. Its subsidiaries include Kongsberg Maritime, Kongsberg Defence and Aerospace, Kongsberg Digital, etc. In terms of business, Kongsberg Maritime is principally engaged in providing integrated products and solutions, such as training courses, hydrodynamic services, ship design services, engineering and service, etc. In 2019, Kongsberg also acquired Rolls-Royce Marine. Of course, it also has fully adopted many connotations of both 'AAWA' and 'Remote and Autonomous Ships—The next steps', with the aim of becoming the world's leading technology company within maritime operations.
(2) Yara International
    Yara International is a Norway-based international chemical company, that is responsible for manufacturing, distributing, and marketing fertilizers and related industrial products. The purpose of its establishment is to be responsible for feeding the world and protecting the earth. In addition, in view of Kongsberg Maritime's leading position in marine technology, in 2018, Yara International and

Kongsberg Maritime launched the world's first emission-free container ship plan. The Yara Birkeland is the first commercial container ship built from the cooperation between Yara International and Kongsberg Maritime. With room on board for 120 TEUs, it will allow air to reduce carbon dioxide (CO_2) emissions by the equivalent of 747 tons between two places by diesel truck per year. In addition, Yara International and Kongsberg Maritime launched its first commercial sailing in 2022. According to company news, these successes mean the company is on track to meet its goal.

It is worth mentioning that, in 2018, Kongsberg Group and Wilhelmsen Group also announced the establishment of the world's first operator of autonomous vessels—Masserly. Among them, Kongsberg is responsible for the development and delivery of whole key enablement of the whole ship, and Wilhelmsen is responsible for the logistics and ship management operations. Therefore, on the basis of the establishment of Masserly, Norwegian Maritime Cluster will still take a key international position in the maritime field.

4. United States of America:
   The purpose of the US Navy developing unmanned vessels is for defense applications. In 2016, the US Defense Advanced Research Projects Agency's (DARPA) Anti-Submarine Warfare Continuous Trail Unmanned Vessel (ACTUV) program developed Sea Hunter and Sea Hawk. Sea Hunter utilizes two diesel engines, with a capability of 27 knots measuring 40 meters in length, and can travel for 19 000 km. Sea Hawk, established in 2021, is an upgraded version of Sea Hunter which displaces that of a 145 long tons trimaran, with the aim to enhance capabilities for naval operations. As we can understand, the development of a whole new unmanned surface ships technology will also be an important initiative in enhancing the US Navy's capabilities. Of course, the potential adversary, China, also has a place in defense applications for unmanned surface ships. Therefore, for the US Navy, it is absolutely essential to provide a more efficient, lethal, and powerful firepower. Although the capabilities and utility of developing unmanned vessels for defense applications remains to be seen, it is undeniable that the US Navy will increasingly focus on developing unmanned vessels in the future.
5. China:
   wIn 2016, China Classification Society (CCS) launched the 'Intelligent Ship Specifications' to keep up with the development of smart ship technology and ensure safety. In 2017, in Shanghai, China revealed the world's first smart ship—Great Intelligence, which has a freight capacity of 38 800 tons, and is a 179-meters-long modified version of the Green Dolphin fuel-efficient bulk carrier. Besides, in addition to the 'intelligent ship' nation from CCS, it is also the world's first smart ship that was awarded 'cybersafe', 'cyber-perform' and 'cyber-maintain' smart-vessel notations by UK Lloyd's Register. In 2018, CCS recognized the service supplier of test field in Zhuhai—The Wanshan

Unmanned Vessel Marine Test Site as the first in Asia and the largest in the world.

Recently, China has increasingly made assertive claims of sovereignty over the sea, thus, it has also been energetically developing an intelligent ship in the defense application field. For example, in 2022, it launched the world's first AI drone carrier which is about 90 meters long, named 'Zhu Hai Yun'. It can carry dozens of aircraft, water vehicles, and equipped submarines drones, as well as coordinate more than 50 autonomous drones. Moreover, there is no denying that China's development of unmanned vessel works will be a focal point of concern around the world in future.

6. Japan:
   In early 2016, the Japan NYK and its group company, Monohakobi Technology Institute (MTI), were selected by the Ministry of Land, Infrastructure, Transport and Tourism (MLIT) to work together and develop '2016 Support Projects for R&D in Advanced Safety Technology of Vessels'. This project was believed to broaden the MASS market. For example, according to data published in 2015 56% of Japan's sailors in their merchant fleet are aged 50 or over, while young people are reluctant to go on board. Thus, the Japanese government is working on possible solutions to address issues of the widely aging population, as well as reduce the dependence on maritime manpower.

   In order to accomplish this goal, since 2020, The Nippon Foundation has been hard at work in developing the MEGURI 2040 program. Fortunately, in 2022, Mikage, the first autonomous container freighter in the world completed a fully autonomous voyage from Tsuruga and Sakai port of 300 kilometers. Mikage is equipped with sensors, RADAR, LiDAR, visual cameras, satellite compass system, AIS data, a remote monitoring system, and a cybersecurity system. Moreover, one of the most challenging works is the autonomous berthing and unberthing. In this work, a drone was utilized to moor with Mikage in the port and assist in pulling a line from the ship to the dock. This is expected to replace crews on board throwing the line from the ship to the dock as an alternative for mooring operations in future.
7. South Korea:
   In 2019, South Korea's Ministry of Oceans and Fisheries (MOF) as well as Ministry of Trade and Industry and Energy (MOTIE) jointly launched the 'Development of Autonomous Ship Technology' for the Korea Autonomous Surface Ship (KASS) Project. The goals include promoting cooperation on core technology for autonomous ships, as well as laying the foundation for commercialization through phased demonstration, with an investment of about 133.3 million US dollars. In this project, KASS cooperated with Hyundai Heavy, Daewoo Shipbuilding, and Samsung Heavy Industries to focus on related standard technologies in order to keep leading the global shipbuilding market in the future. In addition to actively investing in the MASS industry, in June 2019, during the President of South Korea's visit to Norway, both nations reached

cooperation on a MASS-related project. For example, Samsung Heavy Industries signed a memorandum of understanding and cooperation agreement with the DNV GL and Kongsberg Gruppen. The two nations will cooperate to focus on the technology development as well as conduct research related to MASS development work. On the other hand, Avikus, an in-house venture under Hyundai Heavy Industries Group, has developed an advanced voyage support system called Hyundai intelligent Navigation Assistant System (HiNAS), and an intelligent berthing assistance system called Hyundai intelligent berthing Assistant System (HiBAS). Based on the results from the above work, Avikus signed an order contract with two South Korean shipping companies, SK Shipping and Sinokor Merchant Marine in August 2022, which is expected to install HiNAS 2.0 on a total of 23 large ships in 2023. As you will understand, Avikus will significantly enhance the safety and economy of ships with the successful HiNAS 2.0.

3.3 Change of MASS systems

Generally speaking, according to work duties on board, it is divided into two departments: deck department and engineering department. In addition to being responsible for the entire operation of the ship, the deck department is also responsible for the operational safety of the ship at sea and in the port. For example, it must know how the ship waits for a suitable tide or wind to cross obstacles between the sea, shallow waters, rocky areas, and rivers to avoid grounding. On the other hand, the engine department is necessary for the operation, maintenance, and operation of the main engines, power generation and distribution, propulsion systems, and other auxiliary machinery and systems crucial to a ship's operation. In fact, although the work content varies widely between these two departments, the crew in these two departments are trained based on the education regulations of IMO Convention Standards of Training Certifications and Watch keeping (STCW), and at the same time, complete the vocational cultivation experience. Therefore, in addition to the challenging and difficult content of their respective work profile, it is very important that both these departments work in a well-coordinated manner in order to ensure safety, productivity, and efficiency of the ship.

It is predicted that if MASS is to be realized one day, the work of the deck and engineering departments will be thoroughly changed, such as shore duties in the replacement of ship duties. As a result, more and more studies have been purposing some concept of work duties including: Im *et al* [4] suggested that MASS architecture should be divided into two types. One is ship-side and the other is shore-side. Ship-side work allows ships to operate autonomously by collecting and storing relevant information, and then integrating this information for judgement, analysis and processing in a complicated manner. Shore-side work is to receive the real-time status from status information and abnormal information on the ship-side, and monitor the ships or fleets; after that, provide remote control and maintenance functions. Figure 3.2 shows a schematic drawing of the features of ship-side and shore-side.

Understanding MASS operation of ship-side and shore-side is actually quite simple. Figure 3.3 presents the primary control concept of the operation modes

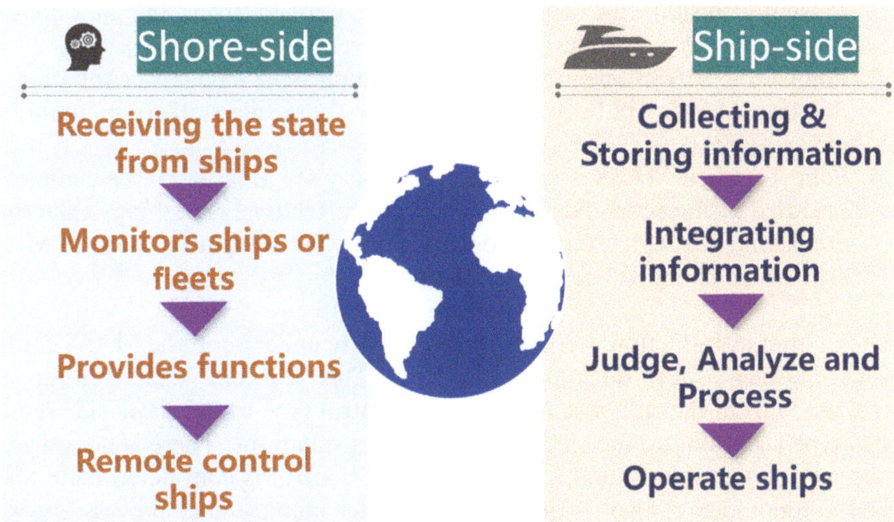

Figure 3.2. (a) Schematic drawing of the features of autonomous ship- and shore-side.

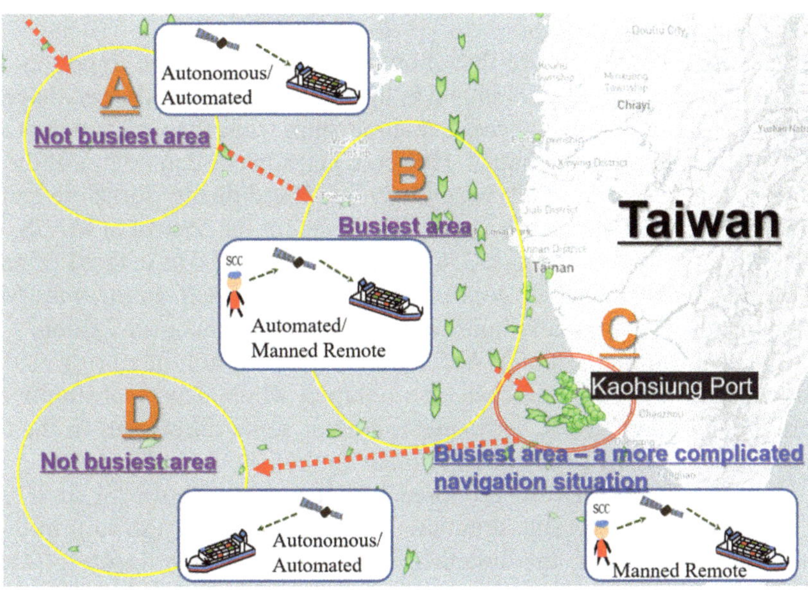

Figure 3.3. Remote control and autonomy of autonomous ships.

between shore-side and ship-side that require humans to operate in special cases, such as:

1. Point A, the autonomous state of the ship is executed from shore-side control when MASS is under autonomous navigation mode. At this time, the shore-side staff can perform relevant tasks in autonomous supervision.
2. Point B, if MASS passes other ships in the busiest area with any possibility of causing collision, the shore-side will be notified from the autonomous systems to carry out relevant collision prevention operations.
3. Point C, if MASS encounters a more complicated navigation situation, at this time MASS will fully be controlled by shore-side staff to maintain safe navigation.
4. Point D, when MASS is navigating in a busy sea area, it can be completely operated by the shore-side or on board crew. Relatively speaking, when ships are passing through the broad open sea area, the navigation mode of MASS can still be changed back to autonomous.

After a brief introduction, we have already clearly understood the MASS control concept between shore-side and ship-side. This concept can transfer the control mode from the ship-side autonomous control type to the shore-side remote control type according to the ships in different sea situations. There is no universal answer, as you can understand, to which type of control is considered better. The type of autonomous control is derived from better management process and safe ways of navigation. The busy waters may implement controls differently than the open sea, yet busy water can still have effective autonomous control according to control conditions. Of course, these control mechanisms can be used quickly and efficiently so that ships can be agile in crafting navigation plans to maintain their shipping competitive advantage.

It can be seen from the above that shore-side control plays an important role in MASS. For shore-side control, as shown in figure 3.4(a), we can also see that a single shore-side control can optimally operate multiple ships while they sail and then achieve the efficiency of multiple ship control. However, ships may often go out of the shore-side control range when they sail, which may increase difficulty for shore control to maintain the ship's position on sea routes. Therefore, for the control of MASS, we can also know from the studies in figure 3.4(b) that MASS can be monitored or remotely controlled by different shore-side controls in different sea areas. In this way, MASS's navigational capability is greatly improved during long-distance navigation.

So far, the control method of MASS development will not only bring changes to the deck department of the ship, but also obtain a whole new opportunity for the engineering department. Next, let's take a look at some differences in the control system of the engineering department. Similar to that shown in figure 3.5, it is 'The engine room system of MASS' proposed by Hogg and Ghosh [5]. It is already known that many unpredictable situations may arise when MASS sails in the open sea for a long time. Thus, the system must maintain a high-degree of available capabilities in monitoring when the main engine of a ship is operating. For example, Autonomous Engine Monitoring and Control (AEMC) is a marine engine

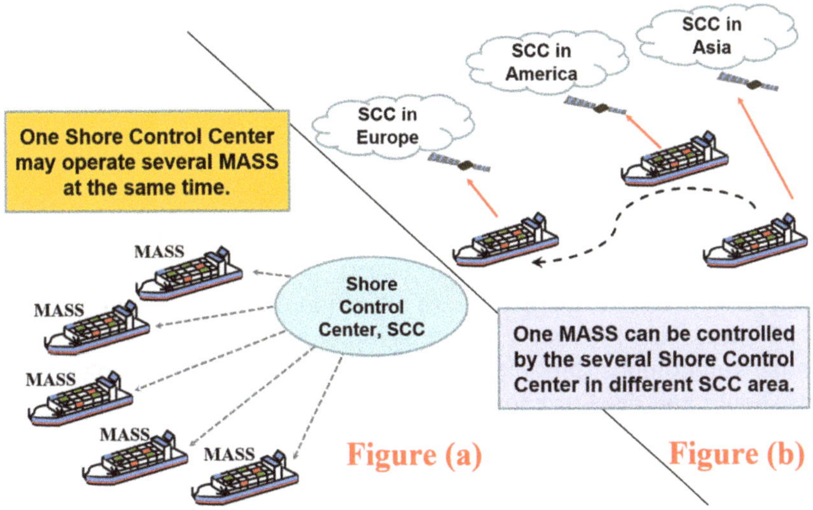

Figure 3.4. Shore-side control mode.

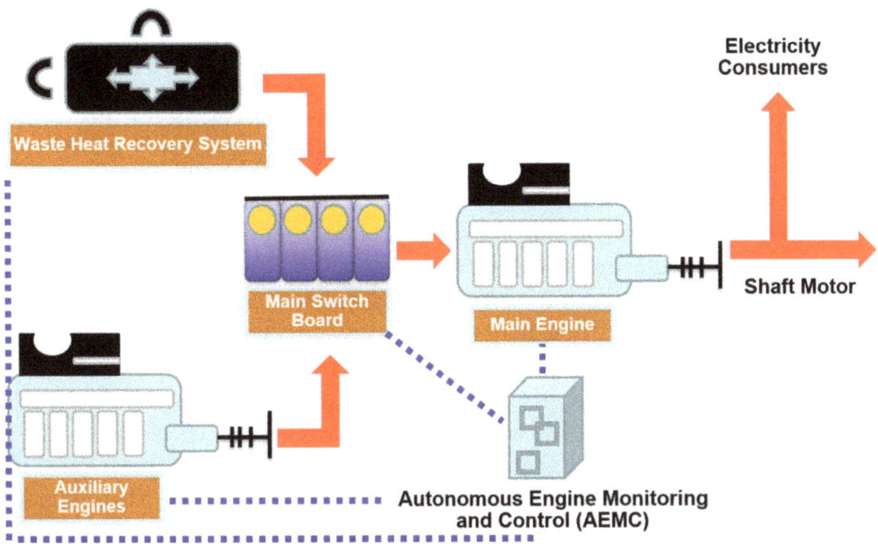

Figure 3.5. The engine room system of autonomous ships.

diagnostic system with advanced fault detection and data processing function. Shore-side personnel can use the results of the diagnosis by AEMC to compile analysis and diagnosis reports.

Indeed, the evolution of MASS has the outline of modern technological development as well as the vision of the future world, which is more related to the transformation of the global transportation and trade system in the future. In the IT explosion era, we have no reason not to develop MASS with related technologies.

Therefore, we look forward to the evolution of MASS, allowing us to have more surprises and very rich imagination about the world, and to welcome a more convenient, more efficient and safer maritime environment in the future.

3.4 Elements of autonomous technologies

Sometimes there's nothing more satisfying than the advancement of technology. Thanks to that, today's transportation is becoming much more convenient than before. Self-driving cars shuttling on the streets, unmanned aerial vehicles (UAVs) delivering packages in thc air, or, best of all, autonomous ships sailing on the open seas. It all sounds like fun, but it would be a pity if you ignore the most important thing in the autonomous ship: the elements of autonomous technologies. So, What elements should you need to know?

Imagine, what technical capabilities a MASS would need to acquire when sailing safely on the open sea or when avoiding collisions between ships and ports as they enter or leave the ports. In fact, to realize MASS sailing, special or different components such as sensor system, control system, decision-making system, etc is needed. In terms of sensor system, MASS also needs to have the ability to sense environmental changes in the ocean just like humans. For example, a camera with visual capabilities (seeing), an Internet of Vessels (IoV) to obtain information (hearing) from the world, as well as Light Detection and Ranging (LiDAR) to sense changes in the outside world (tactile) etc, and the Global Positioning System (GPS) that can provide further guidance and targeted advice. Figure 3.6 shows the most advanced technical devices in autonomous ships.

1. Visual spectrum camera
   In order to provide an approach to obtain high-quality image results of the ship dynamic properties in environmental change, the visual spectrum camera has been utilized to capture more sea state characterization in MASS, such as, wind, storm, heavy rainfall, wave height, and etc. Due to the characteristics of the visual spectrum camera, it can also obtain the clean, sharp, and focused images, even in locations or scenes without bright sunlight. Because it is more sensitive to infrared, its related studies have also been discussing visible infrared imaging technology [6]. Basically, the typical visual elements of MASS include: visual spectrum camera, LiDAR, ultrasonic sensor etc. Specifically, they play different roles in different functions, and have their own unique performance in special situations.

   For a visual spectrum camera in MASS, besides observing each ship's dynamic properties in environmental changes, it is also very important that the sea-side or shore-side control centers monitor the operation of the ship's internal functions. However, it is very difficult to determine and calculate the size and distance of the subject by only using visual methods. Besides, it has also some restrictions from severe weather. For example, visibility distance can be reduced by heavy rainfall, fog, or icy patches of sea. Therefore, by using the sensing element to determine the surroundings is very important work. Considering LiDAR (which we'll talk about later) is a very expensive

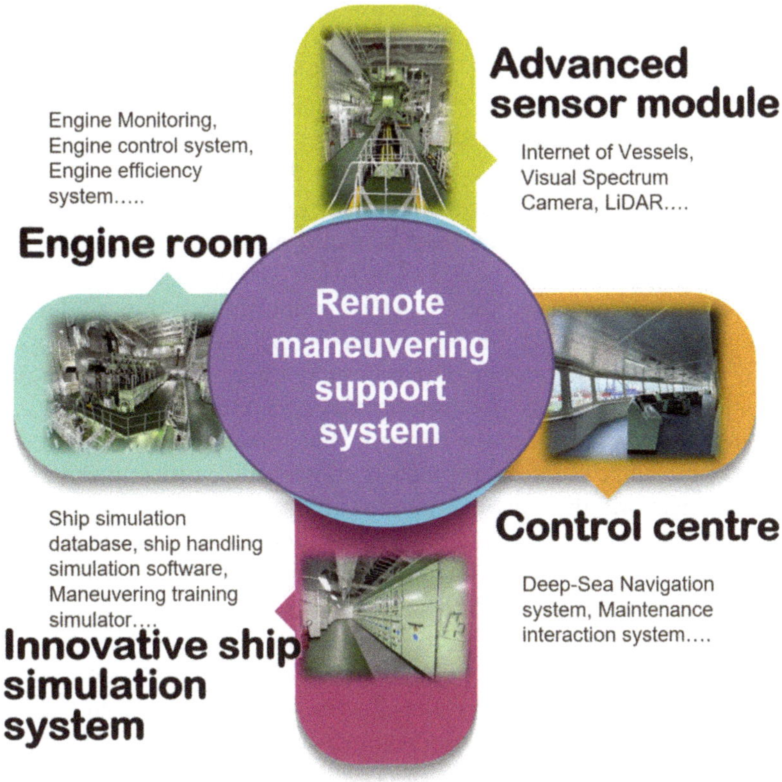

Figure 3.6. The most advanced technical devices in autonomous ships.

sensor and is necessary in creating images of the environment, how to reduce the cost of the image devices and then improve the image quality are attracting many researchers to find a sustainable solution to this issue. In order to find such sustainable solution, the KITTI vision calculation system has commonly used 3D object detection and tracking, especially for the currently automatic driving system technology. For example, it combines visual pictures from multiple angles to create, calculate, and observe the distance and the space of related objects. In addition to this, in order to improve the high-quality image data in MASS, a useful method has been universally utilized in the visual spectrum technologies currently—augmented reality (AR).

AR has been cleverly utilized to enhance natural environments and gather information about the enriching experiences. These technologies could shape MASS maritime transportation in the future. Different from virtual reality, multiple users find themselves living in a completely fabricated world. They also found that augmented reality shows us the man-made world and offers more information. With AR, MASS monitoring has the ability to augment, or add to fabricated experiences by using graphics on a screen. For example, augmented reality monitors can show the essential features of ship identification information, including showing the different color paths in other ships. In the future, almost every MASS will have a high-quality monitor with ample processing power to provide us with a convenient way to augment the reality of MASS.

2. Internet of Vessels; IoV

   I believe everyone has heard of the Internet of Things (IoT), but the term Internet of vessels (IoV) may not be heard often. In fact, the concept of the Internet of vessels is like the IoT. The IoT is for connecting all things through the Internet. In contrast, the IoV is for connecting all things related to ships through the Internet [1]. In order to enable MASS to sail safely on the open seas, it is very important to master and obtain relevant information. In addition, we also know that the operation of much data can be exchanged through the connection of the network, which makes the operation of the ship and the exchange of information more convenient. More importantly, through the ship network, the machine can make correct judgments by itself. That is to say, the possibility of autonomous navigation of ships at sea can be further practiced. At present, there are many shipping companies working to exchange information on the IoV, and their purpose is to gain a leading position in this technology of the IoV. However, there are also many problems to overcome. For example, cyber-attacks [7], cyber privacy, cyber theft, and other hidden concerns. Figure 3.7 shows the framework is built around three main components: maritime services, maritime cloud, and stakeholders.

3. LiDAR

   While obscured, ships and docks are found in the dark with the help of lasers. Scientists studying graphics use laser images to find ships in the open ocean outside of the port. The lasers show that the environment of the port and ocean is bigger than anyone has ever imagined. These lasers, known as LiDAR, can beam from the ship itself while sailing in the area. This technology helps scientists to see through ships, piers, sea levels, and ground surfaces. Besides, LiDAR provides what they have never seen or noticed. In some cases, ships sail past buildings without realizing it. Theoretically, the idea behind LiDAR is to determine, measure, and identify the environment. It is a remote-sensing technique that uses ultraviolet, visible, or near-infrared light to determine objects. Moreover, it mainly adopts 600–1000 nm wavelength bands and then measures the light to returning to the sensor based on the elapsed time.

   Figure 3.8 shows a working schematic diagram of the LiDAR. It can be seen from the working diagrams in figures 3.8(a)–(d) that when an obstacle is placed in the range of the LiDAR detection and scanning, we can obtain the changes in the environment by the laser signals of the LiDAR. For example, the sensors in MASS use LiDAR to detect surrounding objects for high-definition maps. In addition to providing high-definition maps, it can also address the problem of reflection factor for GPS. Therefore, with the help of high-definition map information provided by LiDAR, the positioning of

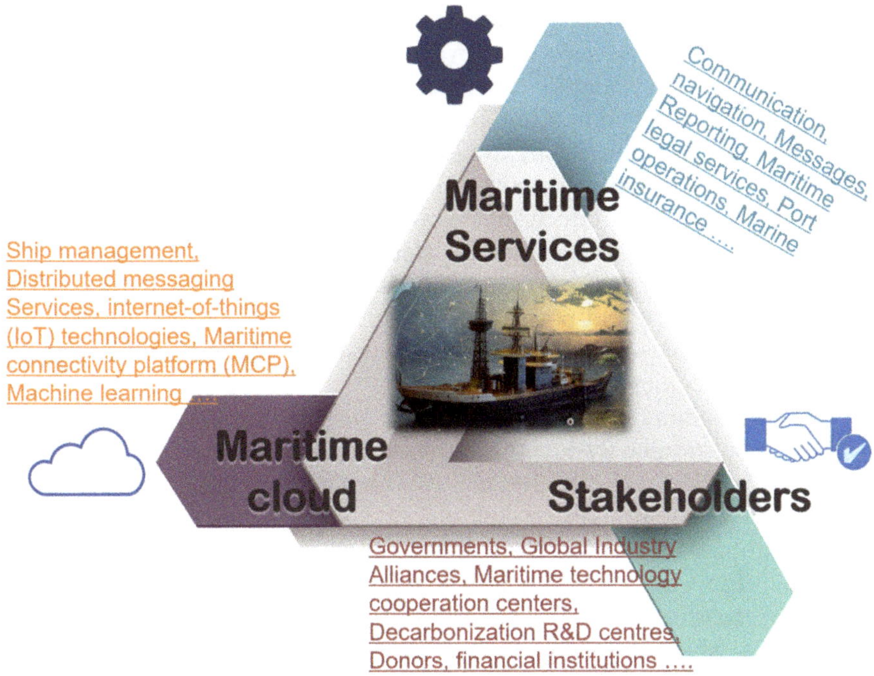

Figure 3.7. The framework is built around three main components: maritime services, maritime cloud, and stakeholders.

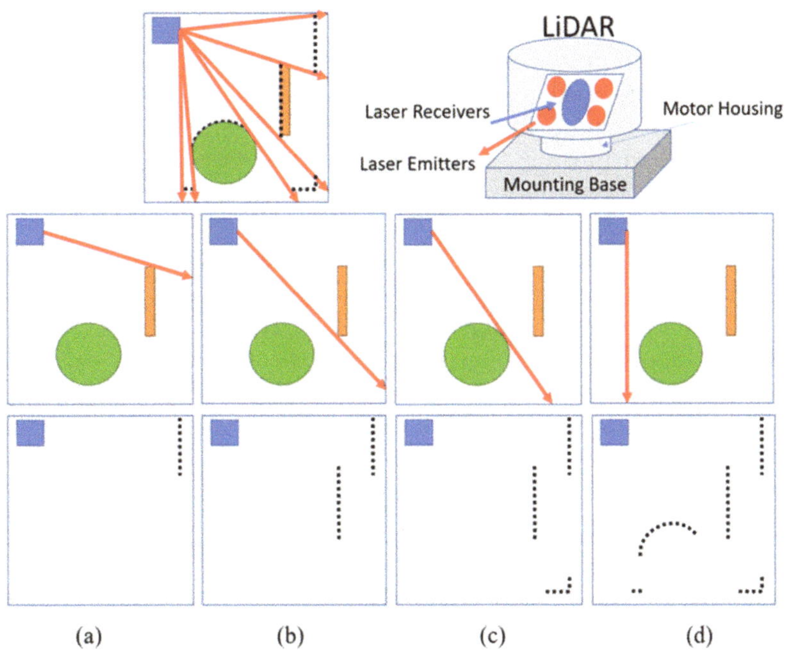

Figure 3.8. The LiDAR working conditions.

MASS can already reach millimeter-level accuracy. Of course, there are still many problems that need to be overcome urgently in the current application of LiDAR. For example, low-precision problems caused by bad weather, high-performance computing, large storage space, and high cost.

As a matter of fact, there are many LiDAR technology companies in the market. For example, Velodyne, Luminar, and Ouster, etc. Among them, not many LiDAR products are engaged in the development of ships. In fact, you can also surf 'The OPAL™' website to see LiDAR in action. Wow! From the website painting, we can clearly understand that the LiDAR is installed on ships and ports, and we can see the range of ports detected by LiDAR. Of course, we also can see the result graphs and comparison graphs of the environment scanned by LiDAR.

4. GPS and communication architecture
Global Positioning System (GPS) has become an integral part of unmanned systems work worldwide, and has also been a hit around the world. You can see it on your cell phone, in self-driving cars, in aerial surveying, in defense applications, and anywhere else you might want to use it in unfamiliar places. The GPS main function is to provide positioning, navigation and timing measurements. At present, GPS consists of a constellation of 24 satellites, which is distributed in six earth-centered orbital planes, each with four satellites, traveling at a speed of 14 000 km h^{-1} and orbiting at 20 000 km above earth.

It mainly uses satellites to work through a technique called trilateration, which means calculated location, velocity and elevation. Trilateration collects signals from satellites to output location information. Although GPS receivers are often confused with triangulation (measure angles), triangulation is used to measure distances, not angles. General speaking, GPS signals coming from satellites down to the ground have to travel through the earth's atmosphere, so they are subject to delays (these delays are random delays which fluctuate). Therefore, it is impossible to precisely measure the situation at any given time.

At present, Differential Global Positioning (DGPS) is used most widely for reducing or eliminating signal deterioration. The current solution is to use an inertial measurement unit (IMU) composed of three-axis gyroscopes and three-direction accelerometers to measure the movement of objects in three-dimensional space.

Another important system related to GPS is the maritime communication system. From the study of [8], it is known that there is a concept of a connection between MASS and satellites. Understanding this concept means that the communication structure of the autonomous systems can be divided into two parts:

1. Responsible for collecting data from different sensors on board:
This includes collision detection and environmental awareness, navigation, safety systems and mechanical operation. Based on the

information from the sensors, the ship itself can make an independent decision, or it can provide a reference for the shore personnel to make a decision.

2. Responsible for the communication between the ship and the outside world:
   Remote operators on the shore can control and monitor the positions of multiple ships at the same time. An autonomous ship can also be able to monitor its own health and environment, and make decisions about sailing based on the information without human supervision.

So far, the above introduced are just a few cases that GPS and communication architecture can do. In the future, there will be more and more applications of GPS and communication architecture available or developed to better improve our lives in various fields. Therefore, if MASS introduces a number of advanced positioning systems and communication technologies, we believe these technologies can truly secure our navigating environment.

3.5 Technical and operational constraints

3.5.1 LiDAR

MASS has a fairly extensive discussion about protecting the environment or reducing the cost to the world over time. In fact, there are more challenges of MASS work than most people realize. As we have done with other novel technologies, we ought to take a step back and determine the ways in which MASS can be beneficial and those in which it is not. Of course, there are plenty of convincing arguments on both sides. Regardless, we can already predict that the transportation mode of MASS will also greatly reduce manpower and transportation costs in the future. However, MASS is still facing many challenges and barriers. For example, if MASS is realized one day, the crew size will must be adjusted [9] and several technologies must be taken to replace crew size change immediately. Next, let us discuss technical constraints in detail and explain how to solve them.

One of the benefits of LiDAR is the high resolution it offers. At the mere touch of a button, we have enough information about environment, distance, and link to multiple system services. It is no surprise that LiDAR has become an indispensable part of MASS. Although LiDAR has many advantages, in the process of LiDAR detecting the environment, the most important factor affecting the accuracy is fog and suspended matter in the air. We can understand that the amount of matter in the air directly affects the measurement accuracy (resolution). Whether it is fog or rain, it is composed of water molecules, and suspended matter in the air is composed of other examples. From the point of view of the distribution density in the air, the probability of the laser light generated by LiDAR colliding with the air will increase during the irradiation process, and the consequent result will affect the accuracy of LiDAR detection of the actual environment. It can be seen that the greater the number of molecules in the LiDAR collision environment, the lower the accuracy of the LiDAR detection of the real environment.

In addition, data processing of LiDAR is also a big challenge. As can be seen from figure 3.8, the more LiDAR laser beams, the more points are generated per second. For example, if a 20 000+ TEU autonomous container ship sails on the open seas in the future [10], do you have any idea how many thousands of data have to be processed at the same time? Generally speaking, the data of reflected light from LiDAR is very primitive data. These data are not just the distance of the reflected object, they must be geometrically transformed through the time factor to generate the coordinator position. In this process, the same action goes through at least multiple spatial function operations. In fact, from the results we obtained in this data processing, in fact, there are more complex operations of coordinate system transformations that can be used for judgment and decision-making. These operations must have quick processing data capabilities to be more efficient.

The biggest key point to success, though, is the cost of LiDAR. These devices are expensive in the market; for example Velodyne offers the price of LiDAR that ranges between \$1000 and \$8000 per unit, or even more. On the other hand, for a 20 000+ TEU container ship, the LiDAR cost will be at least be over \$50 million if the ship has to install more than 1000 LiDARs with the price of around \$10 000 each. Although in affluent nations, prices do not seem like a big deal to many enterprises, numerous market observations have found that too much cost can lead to slow and backward development in autonomous ships. MASS will become so reliant on LiDAR in the future that without the devices, they can barely develop, thus affecting the maritime development.

3.5.2 GPS/IMU

When dealing with GPS errors, a different method is generally used to correct the deviation in accuracy. In theory, the accuracy problems encountered by ships sailing on the open seas will be relatively small, because ships do not have the multi-path reflection problem of high buildings blocking the signal like self-driving cars. The only issue that affects the ship's GPS is bad weather. In addition to this, GPS has a problem with less frequent updates. From the current update frequency, it is about 10 Hz. In order to ensure the safe navigation of MASS, the ability of precise positioning is especially important. Obviously, GPS errors do not increase over time. However, how to strengthen precise positioning has become an important challenge. In order to address the problem of accurate positioning caused by GPS, IMU is generally used as a solution [11–13]. IMUs (inertial measurement units) are primarily sensors used to detect and measure acceleration and rotational motion. Such sensors can be used in automotive stability systems and voyage support system. Fortunately, this sensor is cheaper than LiDAR. Therefore, the combination of IMU and GPS is generally used in the market. Such coordination can improve the problem of update frequency, but it has the disadvantage that each error will increase over time. Therefore, we rely on the IMU for positioning in a very short period of time.

IMUs will have inherent manufacturing defects caused by process factors. For example, when the gyroscope and accelerometer are not spinning or accelerating, the output data will not be zero. In addition, even if the IMU is in motion, errors will still be generated as the usage time accumulates. As a result, the correct location cannot be accurately detected in the end. But more importantly, when the IMU is in motion, the heat source generated by the motion process will also become another serious problem. Therefore, in order to address the problem of precise positioning of the ship by GPS and IMU, Kalman filter (EKF)) seems to have addressed part of the problem [14].

EKF can predict the object position and velocity based on the relevant information. In general, the steps performed by EKF are divided into two stages, the first stage is predicting the time-based position to predict the current position. The second stage is to correct the observed position for further position updates. To put it more simply about the basic concept of EKF, when you wake up in the middle of the night and want to drink a glass of water, you can walk from the bedroom to the kitchen without turning on any lights. In this situation, because you know very well the difference in the relative positions between the kitchen and the bedroom, you can predict your position in the kitchen based on your impressions, and estimate the number of steps taken as well as objects touched by feeling to complete the journey between the bedroom and the kitchen. This is actually the physical concept of EKF. In terms of GPS/IMU, the IMU is the pedometer, and the GPS is your concept of geography. Generally speaking, before we obtain new GPS data, we can only predict the position through the information collected by the IMU. Since the IMU will generate execution errors over time, when the EKF receives GPS data, the two data can be updated and compared to gradually correct the precise and real-time position.

3.5.3 Computer vision-aided systems

It is known that the mass production of LiDAR is limited due to the high cost. However, in the case where GPS/IMU cannot fully grasp the environment, the vision system has become another safety project to assist the navigation of MASS. From many studies, more and more study groups have been developing computer vision-aided systems. Computer vision-aided systems have the advantage of low cost to potentially replace LiDAR in the future. However, in terms of current MASS applications, this does not seem to be as widely used as self-driving cars. Currently, the most mature computer vision-aided system is the KITTI dataset, which was the cooperation between KIT (The Karlsruhe Institute of Technology) and TTIC (Toyota Technological Institute at Chicago) in 2021.

Let's briefly explain what KITTI is. As shown in figure 3.9, we can see a data collection vehicle with many sensors on KITTI's website. Four cameras can be seen from the roof, a pair in color, a pair in black and white, and a Velodyne HDL-64E LiDAR on the roof. There is also a GPS/IMU positioning system. From the footage captured by the four cameras, many pictures are combined into a picture dataset. For example, in the stereo/optical flow dataset, a set of stereo image pairs is captured

Figure 3.9. The KITTI websites: https://www.cvlibs.net/datasets/kitti/.

by two cameras at the same time and at different locations. We can also see from this website's video that the optical flow pictures show the results taken by the same camera at different adjacent time points.

In the KITTI dataset, besides providing all datasets and stereo/optical flow, there are mileage measurement, 3D object detection, object tracking data, as well as road and lane detection datasets. For the KITTI dataset, it can also understand the surrounding 2D images while the vehicle is driving. After processing the KITTI dataset, it is converted into 3D simulated images, receiving the dynamic direction, distance, and object names of the object.

How could the KITTI dataset be used for MASS? It can be imagined that computer vision-aided systems will be used with LiDAR around the ship to provide more high-precision 3D spatial data and obtain better ground truth. For example, there are many related studies using the KITTI database for an autonomous/unmanned ship [15, 16]. In addition, as the application of LiDAR, it can be closer to the actual situation and then replace the images generated by traditional pure computer graphics. According to these advantages, the application of self-driving cars is becoming more and more popular [17], and it is believed that MASS will also use similar devices in the future. However, it is a pity, then, that the development of MASS has not been able to develop at the same time as self-driving cars. Although many researchers are still developing computer vision-aided systems based on images captured by cameras, we expect that computer vision-aided systems will gradually replace LiDAR if the cost of LiDAR cannot be reduced.

It is well known that autonomous ship technologies continue to change our maritime environment, however, the technologies that brought about positive or negative limitations are not always obvious. For example, CNN (convolutional neural network) is also a problem that needs to be addressed and dealt with. CNN is a deep neural network suitable for processing continuous signals generated by images and vision [18, 19]. It can process continuous signals generated by audio, image or video systems to make the picture look more real-time. But in addition to the technologies discussed earlier, the equipment of ship power is also an important issue. For example, MASS will likely be powered by electricity, gradually replacing diesel engines [20]. In this case, the systems of control and operation must still be overcome. From a macroeconomic point of view, all technologies concerned with MASS would have to establish national technical standards. Therefore, the formulation of national technical standards is a very important task for the development of a nation's MASS. This development can set the stage for a nation's development, and it can also implement commercial protection for the nation's enterprises. For example, the technical specifications for MASS established by CCS of the China Society of Ships in 2018. During this period, many nations have also formulated relevant technical specifications.

Indeed, different nations might have different standards, but these standards should serve well anywhere. Thus, once the nations can put them into practice, MASS will be ready to have some very promising development.

3.6 Future trends

Today, in the early 21st century, we find ourselves on the brink of yet another age in the ship's history. Electric ships are set to make the diesel-guzzler obsolete, MASS may soon make human drivers a thing of the past. Nikola Tesla would surely be amazed at how far his dreams have come.

As mentioned in this chapter, it is well known that leading maritime nations of the world, such as the UK, China, Norway, Finland, and Japan, have regarded MASS development as an extremely important project. Even some media reports that the invention of MASS has been a significant turning point in the world's transportation evolution since the internal combustion engine (ICE) was invented. Why do we say that? It is because MASS brings so many great opportunities to the maritime industries based on the maturity of semiconductors and information technologies. However, the sustainable MASS development may still contain some negative factors. For example, the work opportunities for crews are a serious issue. The crew size is expected to decrease, with the percentage of crews estimated to be under 10 people by 2050 for ocean-going ships. The main contributing factor to this is that, thanks to advances in science and technology, fewer and fewer people do not want to work on board, especially in advanced nations. Furthermore, addressing declining birth rates is also now a more serious issue in these nations than before, so fewer people prefer to work on board while working opportunities on land increase. Finally, though traditionally people of developing nations took on the role of working on board, a greater proportion of these people now are willing to work on land. These complicated issues are beyond the scope of this book, so we will not describe them in further detail. Nevertheless, the influence of MASS is still beyond our imagination. Therefore, let's briefly describe these advantages as follows:

Accident prevention: According to statistics, approximately 90% of all maritime accidents are caused by human error [21]. Thus, the realization of MASS will reduce most maritime accidents or casualties caused by a mistake made by crews.

Increasing cargo capacity: MASS no longer needs accommodation and the associated crew systems, such as the crew's living quarters, even the water/air cooling and heating systems, refrigeration systems, and pumps/valves/pipes and bends, etc related to crews. All this will provide more efficient use of ship's limited space requirements.

Crew cost savings: As its name suggests, MASS does not need crews on board anymore. However, there is still a long way to go before it will be completely crewless on board. Before then, MASS will be remotely monitored from a shore Control Centre (SCC), as well as one or two crew members on board on standby.

Reducing ships' fuel consumption: According to related studies, the goal of MASS's propulsion power source will be substituted by fully electric or hybrid systems. This is because the fully electric drive has more advantages of prompt response actions and a small moment of inertia. In contrast, the operation of ICE has large moments of inertia, resulting in delayed response, which means these disadvantages will pose a very serious threat to the safety and security of MASS sailing.

Improving autonomous technology for maritime industries might sound like a useful way to develop the MASS project, indeed, it may soon seem remarkable. That's because more and more nations have encouraged related enterprises to develop autonomous technology as a national key project.

Other recent actions include many multinational enterprises, organizations, institutions, etc are rushing to invest in autonomous technology, hoping to make MASS technology become a leading global technology centered on themselves. These nations believe that since many key technologies and patents are bound to accumulate rapidly in a short time, they have no choice but to try more aggressive activities. Their hope is that when autonomous technology of these actions goes viral, more people will finally realize how important developing the MASS projects is. But not every nation thinks their prerequisite conditions are adequate. While many nations agree that the development of MASS projects is an urgent priority, they also argue that all nations must build sustainable competitive advantage in developing MASS projects according to the development, advantages and limitations of various nations. For example, the author thinks that Taiwan should give full play to their advantages [22], such as semiconductor manufacturing chain, IT, etc, innovative research and development to develop autonomous ship technology.

In any case, internationally, we have also seen huge opportunities and challenges for MASS [23], as what was once shocking will quickly become more eye-opening. By working together, there is no denying that MASS is going to provide a cleaner, more convenient, more efficient, and more comfortable environment in maritime transportation.

References

[1] Koikas G, Papoutsidakis M and Nikitakos N 2019 New technology trends in the design of autonomous ships *Int. J. Comput. Appl.* **178** 4–7

[2] Kretschmann L, Rødseth Ø J, Fuller B S, Noble H, Horahan J and McDowell H. MUNIN D9.3 2015 Quantitative assessment *Technical report*

[3] Poikonen J, Hyvonen M, Kolu A, Jokela T, Tissari J, Paasio A 2016 Remote and autonomous ships—The Next Steps. Technology. Available online: https://www.rolls-royce.com/~/media/Files/R/Rolls-Royce/documents/customers/marine/ship-intel/aawa-whitepaper-210616.pdf (accessed on 1 December 2019)

[4] Im I, Shin D and Jeong J 2018 Components for smart autonomous ship architecture based on intelligent information technology *Procedia Comput. Sci.* **134** 91–8

[5] Hogg T and Ghosh S 2016 Autonomous merchant vessels: examination of factors that impact the effective implementation of unmanned ships *Aust. J. Marit. Ocean Aff.* **8** 206–22

[6] Brando V E, Lovell J L, King E A, Boadle D, Scott R and Schroeder T 2016 The potential of autonomous ship-borne hyperspectral radiometers for the validation of ocean color radiometry data *Remote Sens.* **8** 150

[7] Kavallieratos G, Katsikas S and Gkioulos V 2018 Cyber-attacks against the autonomous ship *Computer Security* (Cham: Springer) pp 20–36

[8] Höyhtyä M, Ojanperä T, Mäkelä J, Ruponen S and Järvensivu P 2017 Integrated 5G satellite-terrestrial systems: use cases for road safety and autonomous ships *Proc. of the 23rd Ka and Broadband Communications Conf. (Trieste, Italy)* 16–9

[9] Kooij C and Hekkenberg R 2020 The effect of autonomous systems on the crew size of ships–a case study *Marit. Policy Manag* **48** 860–76

[10] Bertram V 2020 Technology trends for ships and shipping of tomorrow *Marit. Technol. Res.* **2** 1–18

[11] Liu S, Tang J, Zhang Z and Gaudiot J L 2017 Computer architectures for autonomous driving *Computer* **50** 18–25

[12] Campos D F, Matos A and Pinto A M 2020 Multi-domain mapping for offshore asset inspection using an autonomous surface vehicle *2020 IEEE Int. Conf. on Autonomous Robot Systems and Competitions (ICARSC)* (Piscataway, NJ: IEEE) 221–6

[13] Marques H, Gomes L and Costa A 2017 Emulation of electronic instrumentation devices supporting sailboat's autonomous navigation *2017 IEEE 26th Int. Symp. on Industrial Electronics (ISIE)* (Piscataway, NJ: IEEE) 1255–60

[14] Al Bitar N, Gavrilov A and Khalaf W 2020 Artificial intelligence based methods for accuracy improvement of integrated navigation systems during GNSS signal outages: an analytical overview *Gyroscopy Navig.* **11** 41–58

[15] Omrani E, Mousazadeh H, Omid M, Masouleh M T, Jafarbiglu H, Salmani-Zakaria Y and Kiapei A 2020 Dynamic and static object detection and tracking in an autonomous surface vehicle *Ships Offshore Struct.* **15** 711–21

[16] Li Y, Ma L, Zhong Z, Liu F, Chapman M A, Cao D and Li J 2020 Deep learning for LiDAR point clouds in autonomous driving: a review arXiv:2005.09830v1
[17] Tumasov A V, Tyugin D Y, Porubov D M, Filatov V I and Pinchin A V 2019 . Development of the unmanned vehicle traffic control system with electric drive *IOP Conf. Ser.: Mater. Sci. Eng.* **534** 012030
[18] Gao H, Cheng B, Wang J, Li K, Zhao J and Li D 2018 Object classification using CNN-based fusion of vision and LIDAR in autonomous vehicle environment *IEEE Trans. Ind. Inf.* **14** 4224–31
[19] Lan S, Yu R, Yu G and Davis L S 2019 Modeling local geometric structure of 3D point clouds using Geo-CNN *Proc. of the IEEE Conf. on Computer Vision and Pattern Recognition* 998–1008
[20] Haseltalab A and Negenborn R R 2019 Model predictive maneuvering control and energy management for all-electric autonomous ships *Appl. Energy.* **251** 113308
[21] de la Campa Portela R 2005 Maritime casualties analysis as a tool to improve research about human factors on maritime environment *J. Marit. Res.* **2** 3–18
[22] Cheng H-H and Ouyang K 2020 Development of a strategic policy for unmanned autonomous ships: a study on Taiwan. *Marit. Policy Manag.* **48** 316–30
[23] Felski A and Zwolak K 2020 The ocean-going autonomous ship—challenges and threats *J. Mar. Sci. Eng.* **8** 41

Chapter 4

Impact, development and potential for MASS

Development of MASS will provide economic, social, environmental, and safety benefits to worldwide maritime transport. By developing MASS directly on the goals, MASS will cut down the crews that work on board, which will increase the profits of ship owners as well as allowing them to significantly improve the quality of transport. Of course, MASS is widely limited as well. Not only does MASS technology help provide a convenient source of developmental impacts, it also affects the way we train our crews, shipping safety, and convention implications, etc. In brief, at this time, we are also beginning to fully understand the impact of the MASS industry around the world.

4.1 Crew careers status

When you hear the word 'crews', you might automatically think of those working on a ship, aircraft, or train. Especially for the crew work on board, such as 'maritime crews (marine officers)' guiding ships on the open sea, through dangerous and busy waters, or in and out of a city's port. On the other hand, some 'maritime crews (marine engineers)' work in the engine room of the ships where they must know how to operate every kind of ship equipment, especially for giant cargo ships. These are two of the most important and most difficult jobs in the maritime industry. For these reasons, to become a qualified crew, it takes years of education, training and experience, etc. For appellation in the following chapters, therefore, we replace 'marine crews' with 'crews' to avoid any confusion.

The crew is always the most valuable asset of the ship owner. Moreover, the crew size has also always been an object of concern to the ship owners. However, as is MASS gradually realized, the crew's working livelihood bears the brunt. For example, we discussed earlier about the reduction of crew size from about 200 to the present 20 (in a 20 000+ TEU ship). According to estimates by the International

doi:10.1088/978-0-7503-3151-7ch4

Chamber of Shipping, there is about 1.5 million in the population of crews worldwide currently. It is estimated that this number will be even lower than now when ships are autonomous or fully automated in the future.

From this data collection, we can fully understand and explain the many reasons why the working livelihood of crews has been among the first to bear the brunt. On the other hand, for example, from the views of technological development, technological advancement can reduce the workload of crews on ships, so that crews can have more time to carry out the navigation safely. However, according to related studies, crew working hours, fatigue and health concerns have also always been a rather priority issue [1]. In particular, from much working experience, 90% of the captain's work is dealing with administrative affairs on board per day, which is extremely dangerous for maritime safety. In addition, most people think reducing crew size is for crews' safety.

Nevertheless, according to many studies, more than 85% of maritime accidents are related to human error [2]. In order to reduce the 85% of maritime accidents, there are many studies discussing how to reduce the man-made risk of maritime accidents. For example, studies have investigated the factors influencing maritime safety, including factors related to humans [3, 4], ships [5–7], law [8, 9], environment [10, 11], and management [12–16]. Therefore, as can be seen from the above studies, there are still many aspects to be taken into account in order to ensure maritime safety.

Can reducing the crew size reduce maritime accidents? Maybe this is a rather intuitive idea, but it is not the most suitable reason for developing MASS technology. Although several studies mention that the reason for developing MASS refers to reducing the number of crew to reduce maritime accidents, it may also bring more novel maritime accidents. The author's opinion is the same, and he also believes this to be an extremely negative attitude. For example, in the process of submitting the author's papers, it was found that several scholars think it incorrect to believe the idea that developing MASS can reduce crew errors. Obviously, the author believes that the development process of MASS can bring a better future to people. With such positive thinking in mind, a better and better marine environment will indeed be created. Therefore, the development process of MASS may reduce the related traditional maritime accidents, but it may also generate novel maritime accidents. In any case, we believe that the impact of the novel maritime accidents will be much less than people expected.

On top of that, what new challenges are brought to the crew in the development of MASS? Indeed, an easy challenge that comes to mind is the 'Proficiency', such as the degree one of MASS in IMO:

> *Ship with automated processes and decision support: Seafarers are on board to operate and control shipboard systems and functions. Some operations may be automated and at times be unsupervised but with seafarers on board ready to take control.*

As can be seen from the above, the situation of degree one is the presence of crew on broad. The only difference is that the mechanization, information, automation technology has played a significant role in this kind of ship. Therefore, the operating mode of the overall situation is not much different from the current traditional ships. The only possible behavioral impact of the content of crew work is from the scale of ship automation. Moreover, if the development of MASS further increases, the impact will be more significant. For example, let us further increase degree three of MASS in IMO:

> *Remotely controlled ship without seafarers on board: The ship is controlled and operated from another location. There are no seafarers on board.*

At this time, there is no crew on board, and the ship control is made from the shore. This situation raises so many questions in our minds. For example, some people may wonder what kind of jobs have disappeared for the crew? Has the crew's job transferred to the shore? Is the crew still called the crew at this time? This series of questions is not easy to answer. It needs more time to respond and define. While it may not be necessary to change the definition of crew from current standard form (such as, STCW), we think the core of this issue lies in the contribution of the crew's future development of MASS.

Simply speaking, crew work on board has not disappeared, only the essence of the work has changed. For example, Ahvenjärvi [17] proposed that crew training and education have a significant impact on improving maritime safety management for MASS, which involves creating new maritime accidents, while Porath *et al* [18] proposed that the relationship between MASS and human error is related to the degree of autonomy, which leads to change in the essence in related work. Therefore, working on board needs to be extremely professional and experiential. At the same time, regarding navigation rules, collision avoidance, climate by judgment, it is not easy for a person without navigational expertise to be responsible for these tasks. On the other hand, in the future, crews also may require the ability to work on the shore. By the way, under Rolls-Royce's autonomous ship plan, one or two crew members will still work on board to deal with possible, unexpected marine problems as well as provide a security option for MASS at sea. It seems that the crew of the future may not only have a major profession in navigation or marine engineering like current crew members.

Perhaps these new questions will not lead to more standard answers. However, based on the author's experience, these complicated problems may involve more 'commercial interests', such as insurance, a financial product by insurance companies to protect ship property from the risk of loss or damage. In 1910, the 'International Convention for the Unification of Certain Rules for Collisions of Ships' adopted by IMO dealt with the attribution of liability for collisions of ships. According to this convention, insurance companies make various level-types proportions, ratios, percentages, or fractions according to accidents causing deaths and injuries. In terms of level-types, this proportional reasoning involves complicated commercial interests. Simply put, if two MASS have an accident, does the responsibility belong to ship A or ship B? Whether the responsibility for the ship belongs to the ship, the personnel on the shore, or even the system developer, etc, there are still had many specification challenges that need to be clarified and addressed. As we can understand, due to most of the collisions or accidents of ships involving international regulations, thus, a relevant ship's insurance problems are difficult to handle for an insurance company or a nation alone. Whether the crew is still called a crew or not, or whether human error is still caused by humans, for example, the international regulations still need to be further explained and defined through various governments of nations and IMO. Comfortingly, we already know that these issues are still being discussed. In fact, we do believe that with more and more questions being discussed, the answers will be more clear eventually.

4.2 Maritime industry shocks

The maritime industry is a significant global industrial supply chain. Its product process, operations management, and productivity have always played a key role around the world. In the previous chapters, we have already realized that many technologies and management, such as Internet of Vessels (IoV), shipbuilding, blockchain, and the intelligence of port, etc, all require a high level of integration and development capability to realize MASS performance. Among them, the shipbuilding industry will also play an important role in the MASS development process. Therefore, in this section, let us focus briefly on how MASS development impacts the shipbuilding industry.

In the complex process of shipbuilding, the target of each stage has a different materiality technological threshold for creating and delivering products. For example, with regard to professional talent, shipbuilding production planning requires professional backgrounds in marine design engineers, mechanical engineering, marine engine engineering, shipbuilding engineering, as well as having the responsibility to handle several shipbuilding professional projects independently. Actually, it also reveals the importance of professional technology integration in different fields. Therefore, in order to respond swiftly to changes in the structure of maritime industries, in addition to the modification of the various professional technological methods, the most important thing is talent integration and demands. Of course, due to talent being the most important factor to be considered in shipbuilding industries, the integration of professional work in the shipbuilding industry is a pressing matter that needs to be addressed.

According to the manufacturing, production process of the shipbuilding industry and the relationship between technologies and talents, generally, it can be divided into four stages: ship design, system equipment, construction and assembly, and ship inspection. Thus, let us see what happens next.

4.2.1 The ship's characteristics design

The ship characteristics design process includes interaction between ship stability, vibratory motion, minimum powering, strength action, maneuverability, seakeeping, and other general arrangements. The main talents are required to have the background of fluid mechanics, material mechanics, electrical engineering, engineering mechanics, etc. Nevertheless, whether MASS can achieve navigation at the lower wave-making resistance coefficient in the future, all rely on ship design talents. In particular, MASS with streamlined structures increases valuable cargo space while reducing crew space. Such ship spatial planning is very important in navigation work.

4.2.2 System equipment

Equipment for traditional ships includes fuel, lubricating oil, seawater, daily water, steam, cooling, and propulsion systems, etc. If MASS is fully developed one day, the scale of the crew's life support system will gradually shrink due to the fact that the autonomous ship may only carry one crew member or even none. In contrast, marine engineering, control systems, measuring instrumentation systems, etc inevitably need large-scale transformational changes. For example, the marine engine system is responsible for propelling ships. In the future, if the hybrid power systems or even fully-electric systems are realized for autonomous ship's propulsion, there will be a substantial increase in demand for systems engineering talents for the ship. Therefore, not only is the demand for system engineering talents increasing, but the development of MASS must rely on work division of many professional fields to greatly reduce technical barriers, so as to enable MASS to successfully sail on the open seas.

4.2.3 Construction and assembly

The hull is the main body of a ship construction. It is the part of shipbuilding engineering that includes ship construction and assembly engineering, such as installation, assembly, and restoration, rust removal and coating works, etc. In terms of professional talent, the ship construction process requires the production management and production engineers, etc to propose high-quality management plans of production scheduling, standards, and planning. Therefore, these engineering positions typically require a relevant field that is suitable for the job, such as materials engineering, mechanical engineering, industrial engineering, naval architects, and marine engineers. Thus, in order to realize MASS sailing on the open sea, the quality of integrating high-tech ships will be a critical task regarding the relevant projects and budget considerations. In particular, if MASS is popularized in the future, how to communicate effectively with customers about ship construction, assembly projects and budget will also be an important task.

4.2.4 Ship's verification

In the process of traditional shipbuilding, a ship must be verified by a ship classification society, such as Nippon Kaiji Kyokai (NK), Lloyd's Register (LR), Det Norske Veritas (DNV), American Bureau of Shipping (ABS), China Classification Society (CCS), etc. After passing the complete verification, the ship will be able to sail on the open sea. Despite the many technologies involved in the process of ship construction, a lot of verification work is also required to be dealt with, such as electrical equipment testing, programmable electronic system testing, control and instrumentation equipment testing, and hull structure testing, etc.

In addition, verification of the environment is also essential. For example, the control of Ballast Water Treatment System (BWTS) or sulfur oxide (SO_2) emission will increasingly focus on environmental risks and reduce vulnerability hazards due to the IMO's strong emphasis on avoiding pollution from ships. It is probably quite easy to estimate some roles in these fields, such as environmental engineering, energy engineering or chemical engineering, etc. In addition to these talents, it is still necessary to rely on experts with practical experience to assist in verification, such as the captain, chief engineer or marine pilot, etc. It can be seen from the above, for the realization of autonomous ship related works, that an integrated engineer may also be required to complete the shipment verification work in the future.

As explained above, the foundation of the traditional shipbuilding industry can create more opportunities to support MASS development and implementation for MASS prosperity. Although the impact of MASS may face a lot of different challenges and obstacles, the main sustainable MASS development still depends mainly on the success of the integration works of professionals in the future. In any case, MASS development will also continue to bring a new and improved opportunity to the shipbuilding industry, and then support more unimaginable pleasures in this field.

4.3 The influence of port transportation

As can be seen from the above sections, companies are trying to do their part to improve the environment for maritime transport and the port logistics sector. They are changing port facilities, services, and shipping methods. They are making them friendlier to the MASS environment. In fact, many companies have also shown that the impact of MASS on maritime industries is not only limited to the crew, of course, it also involves things other than the crew and the ship. Now, let us take a look at how maritime transport works and the impact of MASS.

4.3.1 Wireless call

Before traditional ships prepare to enter the port, the communication equipment must be used as a way of transmitting information so as to contact the port and then berth. For example, for MASS, calls and information exchange depend on communication between control shore center (CSC) staff and harbor management. Ideally, MASS should complete the voyage plan prior to departure, and complete the navigation and port of call work under reliable communication. In the future, it is conceivable that MASS communication operations may be carried out by shore-based personnel and port units [18].

4.3.2 The pilot service

When a ship enters a port that is not familiar with climate, hydrology and related port facilities, local experienced pilots must guide the ship. One question to discuss is whether pilot jobs will disappear with the emergence of MASS? This issue may be addressed in two respects. First, the work of pilots may disappear. The reason is very simple. When talking about AI in the previous chapter, we can implant the experience of the pilot into the AI of MASS, so that MASS can complete the guiding work of the ship through the operation of AI. In this process, there is absolutely no need for the pilot. The second aspect is that the pilots are still needed. Because of the unpredictable changes in climate and hydrology, or the bottleneck in the development of AI technology, the pilot can guide the work of MASS from the side. Whether pilots will disappear or not, it is certain that the pilot's work will be affected.

4.3.3 The work of the tugboat

Unmanned tugboats are also one of the projects of ship development. For example, Abu Dhabi Ports company is developing the world's first fully unmanned commercial tugboats. When the ship is ready to berth, it usually needs to use the tugboat's own auxiliary propulsion as a tool for berthing. However, in order to complete the berth operation more quickly and safely, tugboats are undoubtedly an effective tool to help berth. However, the work of tugboats is actually extremely dangerous. The process of using tugboats to hold a large ship or pulling ships with ropes is also a large risk to tugboat work. Fortunately, the realization of unmanned tugboats will reduce this risk, but it may also impact the job opportunities of tugboat crews.

4.3.4 The mooring and unmooring works

The mooring and unmooring works in the wharf are usually done by dockers. The realization of MASS may force dockers to face the possibility of losing their jobs. For example, according to YARA's MASS design, it can be found that the emergence of MASS will make the wharf to become automated and unmanned. We can predict that in the future, the mooring and unmooring works, cargo handling, or supply operations will gradually be replaced by automation. However, in view of the high population density of wharf projects, governments of all nations should seriously consider how to reduce the impact on the dockers and develop relevant, appropriate supportive measures. Especially, considering that each requires different change strategies, plans and degrees of employee engagement. To be clear, in a situation where job opportunities have a tremendous impact on dockers, the transformation of the dockers on the wharf will be a second to none.

4.3.5 The cargo handling

Loading and unloading workers will also be affected by job losses. Take a container ship as an example, it usually takes about two minutes to load and unload a container. For example, loading and unloading operation basically requires the joint completion of the ship's officer, the crane operator, the wharf truck driver and the shore container dispatcher. This work is critical to the smooth delivery of goods to consumers. Therefore, this work is particularly important. The realization of MASS in the future will affect the entire operating system. For example, it can be found from the designs of YARA and Wärtsilä that unmanned and automated ports are bound to be realized, and reducing this impact has become another important development topic. It is understood that China's Qingdao Port has already set up relevant tests of unmanned docks. Therefore, before the realization of MASS in the future, it is a very important economic issue to develop relevant supporting facilities to reduce the impact on loading and unloading and terminal staff.

Although the realization of MASS has a great impact on the maritime industries, it is an inevitable phenomenon. From these impacts, it can be seen that whether it is pilots, tugboat crews or dockers, the relevant work content and job types is bound to vary according to economic, technological, environment conditions, etc. More importantly, these changes require the cooperation of the government, enterprises, and dockers to minimize the impact. Therefore, in the future, on the road to MASS development, there may still be some impacts that are not easy to address or have not yet been discovered. But in any case, these various aspects of the impact on MASS are indeed worthy of our serious reflection.

4.4 Safety and security in MASS

News and reports about autonomous unmanned technology often emphasize convenience and safety to us. Such news reveals that one day, these technologies may actually replace traditional merchant ships. However, in the real world, some people may still be concerned about autonomous technology or AI because these intelligentized machines may raise some concerns about the system stability and reliability which could also cause mistakes to our safety.

The best way is to learn how to be safe. Besides, even the most reliable technology or equipment can still go wrong. That is why MASS should make sure that the sensor, driver system, control system and even IoV are working normally and prevent it from being destroyed. MASS should know when and where a dangerous environment exists, and will even notice itself if the autonomous system is deteriorating. Figure 4.1 shows the ship's potential and realistic threats. Evensen [19] thinks that every shipping company should conduct a threat assessment to recognize the potential and actual threats they may be faced with. Thus, along with the ship's complexity that continues to increase, with that comes an increasing threat of attack of the surface.

Also, it should never ignore some signs of attack. The chances are that if anything happens, MASS should able to correct judgments. For example, if a visual spectrum camera fails, LiDARs will help the visual spectrum camera decrease the amount of time it takes to get aid in an emergency, and the visual spectrum camera will be doing the same for LiDARs. In addition, according to the author's experience in transportation safety works, the principles and practices of transport operation management should be: safety first, reduce traffic congestion, and finally save energy. Obviously, safety is a primary consideration in this major management sequence principle. Furthermore, in order to maintain a safe MASS's development, we should first seek to understand what work problems we have encountered, will encounter, or have encountered in the past during navigation. The following items

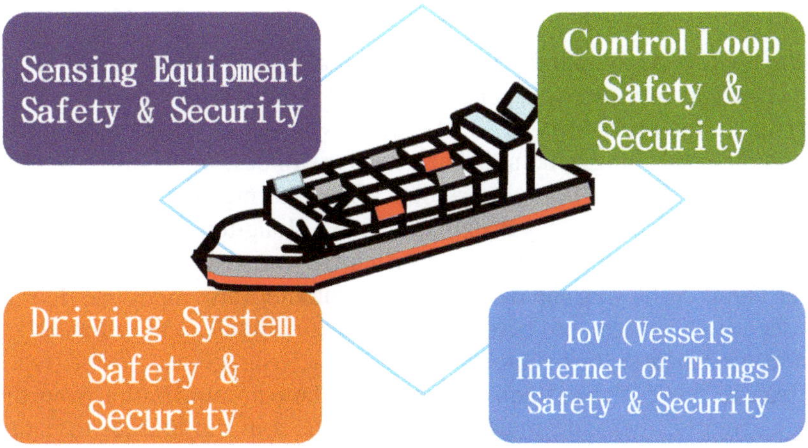

Figure 4.1. The ship's potential and realistic threats.

provide a study issue for navigation of the autonomous ship, and when you have completed reading, you should be able to know.

4.4.1 Sensing equipment safety

If MASS were to be attacked or intercepted by environmental or outlaw factors, the most direct way is to attack the hull and sensor of the ship. The environment and weather are obvious examples. Will MASS be forced out of navigation if a storm suddenly hits the sensor or hull? Strong winds from typhoons or hurricanes in busy water are especially harmful to MASS sailing. Another example is the collision of maritime accident cases. MASS can tell when other ships are approaching as their sensor detects changes in the environment. As a result, autonomous ships can deviate earlier to avoid the collision.

On the other hand, since sensor error is dangerous for MASS, it is not surprising that damage by human action can be even worse. As we mentioned for optical cameras in the previous chapters, if someone lays smoke around the autonomous ship, or someone uses the environment to interfere with the optical cameras, then the autonomous ship will be easily attacked. In addition, someone might use reflective devices to interfere with the LiDAR so that it cannot detect the surrounding environment with high accuracy, resolution, and length detection, which results in the return of false information. In this situation, when the LiDAR is interfered with, the ship will easily be in danger when navigating. Of course, such examples are also used in GPS or IMU.

Thankfully, many sensors are built inside MASS to help them stay safe in dangerous situations. Unlike traditional ships, autonomous ships need many different sensors all the time, so they are not subject to one sensor error. For the functions of autonomous ship sensors, besides being able to detect the environment in their various sensors to maintain safety, each MASS sensor calibrates their signals to others to prepare for dangerous situations and ensure safety.

4.4.2 Driving system safety

Attacking the driving system safety of MASS is also a significant threat. The attack method is an interruption or suspension of the driving system workflow. In general, the current autonomous ship drive operating systems use ROS (robot operating system) as the controlling structure. The attacker can destroy the node work in ROS as a target, and make a heavy workload affect the memory performance in the computer. Finally, the memory is damaged due to long-term consumption, and the node work is discontinued, eventually causing the entire drive system to crash. On the other hand, destroying the drive system is making false messages. This method causes the drive system to perform erroneous actions, and ultimately causes continuous destruction to the autonomous ship.

The way to prevent the attack is to encrypt the communication from the drive system. Currently, there are many system suppliers in the market to encrypt the work of ROS nodes. For instance, a common method is to use DES (data encryption standard) to encrypt. However, encryption does have an impact on the drive system. In general, the more node works there are in the system, the more encryption there is. when the number of encryptions increases, the processing time will increase as well. This will bring additional workload to the system and increase the security risk.

4.4.3 Control loop safety

Advancement in MASS technology has made our maritime lives far easier, but also made it easy for others to control your MASS. Here are some common methods that are used by hackers and how we can deal with them.

It is likely that you may rely on a ship area network (SAN) like the controller area network (CAN) of an unmanned car to do everyday tasks. SAN busbars appear in metallic bars in a switchgear panel used to connect all electromechanical equipment at the ship's electrical switchyards and distribute them to the outgoing feeders. If hackers attack this device, they may attack the OBD-ll (on-board diagnostic) first, then control your personal autonomous ship data. Thus, it is important to lock the communication protocol of the device with a strong password. For example, your password should use a mix of letters, numbers, and symbols to make it as difficult as possible to guess. There are other ways to protect MASS data. In the future, most devices will come with facial recognition or fingerprint scanner features. These ensure that only the owner of the device is able to use it. Sometimes hackers can defeat these features, so remember to update the function. This will allow you to locate a stolen device, lock it remotely, and sometimes even delete all the data on it.

4.4.4 Internet of Vessels safety

IoV can provide you with the ship information you need through operating systems and communications. IoV always needs to update the operating system or software, as new versions of operating systems can fix problems in previous ones. In addition, when you install some updated IoV software, they will often ask for permission to access functions such as devices, recorder, or GPS. You have to make sure that if you do not think they need this access, do not give it to them.

Of all the ways IoV information or data may be used, though, the most common is through the safety of V2X (vessels-to-everything). This is an online way to protect your IoV information or the links you click. If you find your ship information in an unsafe situation, remember the following things. First, never give out your shipping information in an online message no matter who is asking. Second, never click on a link that looks fishy. After all, protecting your shipment data may seem like difficult work. But remember, all the effort is worthwhile as long as you can prevent your ship information from being stolen.

4.5 Legal implications of MASS

International maritime shipping is a process of importing and exporting goods between different nations via the ocean. Many nations are trying their best to stay on the safe side of international marine convention. Most of the time, this is relatively easy—do not break the law, and obey the maritime traffic regulations. But the law is long and complicated—in IMO, the International Convention for the Safety of Life at Sea (SOLAS) alone includes articles setting out general obligations, amendment procedure, etc, followed by an annex divided into 14 chapters. As a result, there are plenty of seemingly harmless things that are, according to the law, being punished for. Say, for example, when ships are sailing at sea, the laws involved include United Nations Convention on the Law of the Sea (UNCLOS), International Convention for the Safety of Life at Sea (SOLAS), International Convention on Standards of Training, Certification, and Watchkeeping for Seafarers (STCW), International Convention for the Prevention of Pollution from Ships (MARPOL), and Convention On the International Regulations for Preventing Collisions at Sea (COLREGs), etc. So, with these regulations and limitation acts, it will allow crews, ships, goods to reduce the risk in transportation and enhance maritime safety.

When it comes to regulations for MASS, our understanding is that it is very broad. For example, future changes will not only be limited to crews or ships, but will also affect cargo, the transportation environment and port management. To ensure that our discussion is more focused, this section will discuss the legal implications of MASS using the SOLAS convention as an example. So, let us go through the relevant points one by one.

Chapter one of the SOLAS Convention is the general provisions, which stipulate the governments of the nations whose ships have the right to fly the flag. Although MASS have no crew, they are still ships. Therefore, MASS are also regulated.

Chapter two is the regulation of 'Construction Structure, Subdivision and Stability, Machinery and Electrical Installations'. If MASS is to operate on the open seas in the future, some of the relevant regulations need to be reviewed. For example, this chapter involves the special provisions in '5-1 providing stability information to the captain', which may involve contact and positioning between the CSC of an autonomous ship and the crew on board for other related matters.

Chapter three regulates 'life-saving appliances and arrangements'. When an autonomous ship sails on the open sea, it is very important to consider the life-saving regulations between the crew and the passengers, or what kind of survival and operational skills the crew on board of the autonomous ship (Degree 1 or 2) must possess.

Chapter four regulates 'radio communications'. This chapter states that the crew aboard the ship must have effective and safe radio communication capability. In the case of MASS, if this requirement appears unreasonable in only Degrees 3 or 4, a more feasible way is that shore-based personnel take on the associated work. However, this specification remains open to debate.

Chapter five states that 'safety of navigation' means that all ships must be equipped with an adequate and efficient workforce and requires States Parties to have clear legal standards. Based on that specification, these regulations have a limited degree of specification for MASS. That said, it is not totally prohibited to maneuver without a crew on board a ship. As long as the ship has sufficient crew and is assisted by high-tech and specific capabilities, it may be regarded as having appropriate navigational capabilities. In this regard, many nations have developed associated technologies to replace regulatory restrictions. The aim is to resolve the possibility of the autonomous ship operating on the open seas in a safe and unmistakable manner.

Chapter six regulates 'carriage of cargoes and oil fuels'. This regulates operational requirements in relation to the safe operation of the goods. This chapter is similar to chapter seven 'carriage of dangerous goods', which emphasizes that crews must pay attention to the carriage of goods at all times during the shipping. Particularly in the transport of dangerous goods, in addition to requiring MASS to comply with the provisions of the SOLAS Convention, they can also meet specifications in International Maritime Dangerous Goods Code (IMDG Code), which will be a major test for future regulatory work.

Chapters six through nine also discuss a large number of regulations dealing with the relationship between the ship and the crew. This includes management and safety issues as well. For example, chapter six addresses maritime safety, maritime security, and crew duties. This means that the manner in which the crew avoids the threat of pirates and maintains the safety of the ship, crew and cargo is especially important.

Especially when MASS is operating on the open seas, the relevant equipment specifications must cover chapter two, as the old saying goes, 'A small leak will sink the great ship'. Therefore, if an autonomous ship is attacked, how the Convention requires the ship to have sufficient security alarm systems and mechanisms is also an issue worth discussing.

So, while many nations pride themselves on being law-abiding citizens, with the sheer number of multi-faceted conventions on the crews, ships and environment, etc, who knows how many mistakes we might unconsciously be committing. In addition, due to the wide levels of regulation involved, it is inevitable for the design of MASS to cooperate with the need to modify the international conventions in the future. Simply put, we must consider all possible combinations of technique and convention. In this way, the impact of MASS convention will be minimized.

4.6 The post-COVID-19 pandemic era

Containers ships full of goods speed past each other in the sea, delivering MASS clicks back and forth on a work day, like dolphins. It sounds like a scene from a dolphin show in the ocean park, but due to COVID-19, voices for realizing MASS development trend have come much sooner than expected.

During the current pandemic, the need to transport goods by crew contact has caused many companies to push forward with MASS delivery projects. And while some such services were tested in many nations before the pandemic, in the recent post-pandemic period, MASS development has been popping up all over the world. Ocean Infinity, for example, is expanding its Armada fleet of autonomous surface ships to 17 robotic ships over the next two years via a new build program with ship builder VARD and a recent contract for the supply of mooring and rescue boats with Palfinger. On the other hand, in 2021, large companies such as Norway-based Kongsberg Maritime signed an agreement with the Institute of Marine Research (IMR) for the delivery of four autonomous ships. In fact, this order covers two Kongsberg Maritime Sounder USVs (unmanned surface vehicles) and two Kongsberg AUVs (autonomous underwater vehicles).

Besides, the future of autonomous technology delivery promises many benefits to customers and enterprises alike. MASS, for example, can easily sail over oceans and even canals, thus achieving many safe delivery projects. It can also finish the mission that crews try to accomplish. There are, however, still many possible problems to address before this super convenient future becomes a reality. While delivery of MASS may work well in the sea during a pandemic lockdown, it may likely perform not as well in crowded port areas. As a result, special routes for MASS delivery would likely have to be put in place, which means some major changes to our maritime environments. MASS delivery may indeed be proved useful during the pandemic period. The challenge is: can MASS continuously thrive in a post-pandemic world?

4.7 Future trends

Developing MASS is comparatively an environmental choice, because it can decrease environmental impacts that result from transportation and contamination. This means that they use fewer resources than the commercial maritime industry nowadays, which requires more fuel consumption, human resource, and marine risk.

Indeed, risks at sea not only have a tremendous impact on the development of MASS, but also bring many challenges. In this chapter, it is found that the impact, development and potential of MASS involve a wide range of aspects. For example, we mention the crew's work, the maritime industry transformation, the differences between the novel maritime transport mode and the traditional model, the safety issues related to the legislation and crews, as well as the impact of the conventions on MASS. Apparently, these are all key issues that need to be resolved urgently. Therefore, in simple terms, these impact issues can be briefly summarized as the impact of the crew, ship, environment, regulations, and management in this chapter. The impact of MASS can be on on ethics, emotions, or people's insistence on justice. On the other hand, with ethical issue, readers can fully recognize that the autonomous ship's collision program will determine whether it is self-harm or harming others. For these pending issues, we already ensure that it will generate more instituting conflict issues between commercial interests and security concerns in the future.

In order to address the above-mentioned impact, we should set out from the challenges of the basic issues. For example, lack of basic facilities in ports, network transmission delay, operations in severe weather conditions, lack of clarity in accident liability, network security risks, privacy policy, as well as final legally acceptable definition, etc. As can be seen from the above, these challenges involve the construction level, technical level, management level, legislative level, etc. In particular, when many researchers investigating these priority issues based on the collection of many references as well as the results that have aroused extensive international discussions, it is generally believed that the most important challenge should be the 'international legislative norms'. This reason is not actually hard to understand. In order to realize MASS sailing on the open seas, MASS must have authorized international laws to operate on the open seas. For example, we know that a self-driving car can drive freely on its own domestic roads. However, sea areas are different from roads. Self-driving cars on roads only involve domestic regulations. Not only is MASS restricted to domestic regulations in the territorial sea, it is also restricted to international regulations as well.

Moreover, MASS sailing on open seas has long been a hot topic at the IMO recent years. Although IMO and many nations have had several discussions about the direction of MASS development, if MASS is to sail on open seas, rules of conventions must be resolved and emphasized as soon as possible. In fact, this is an issue that many people are concerned about. Right now, this matter is still stirring up controversies. In any case, we believe that the software and hardware of MASS technical issues will not be too difficult to address. However, when MASS involves human emotions, cognition, and ethical issues, the method that must be resolved relies on more MASS applications, experience, and examples. Finally, we believe that these will be the basis for solving the above issues in the future. Therefore, let us expect that this series of issues can be resolved as soon as possible, and then gather here to welcome the arrival of the MASS era.

References

[1] Baumler R 2020 Working time limits at sea, a hundred-year construction *Mar. Policy* **121** 104101

[2] de la Campa Portela R 2005 Maritime casualties analysis as a tool to improve research about human factors on maritime environment *J. Marit. Res.* **2** 3–18

[3] Chauvin C, Lardjane S, Morel G, Clostermann J P and Langard B 2013 Human and organisational factors in maritime accidents: analysis of collisions at sea using the HFACS *Accid. Anal. Prev.* **59** 26–37

[4] Hetherington C, Flin R and Mearns K 2006 Safety in shipping: the human element *J. Safety Res.* **37** 401–11

[5] David M, Perkovič M, Suban V and Gollasch S 2012 A generic ballast water discharge assessment model as a decision supporting tool in ballast water management *Decis. Support Syst.* **53** 175–85

[6] Pazouki K, Forbes N, Norman R A and Woodward M D 2018 Investigation on the impact of human-automation interaction in maritime operations *Ocean Eng.* **153** 297–304

[7] Le Tixerant M, Le Guyader D, Gourmelon F and Queffelec B 2018 How can Automatic Identification System (AIS) data be used for maritime spatial planning? *Ocean Coast Manag.* **166** 18–30

[8] Knudsen O F and Hassler B 2011 IMO legislation and its implementation: accident risk, vessel deficiencies and national administrative practices *Mar. Policy* **35** 201–7

[9] Schröder-Hinrichs J U, Hollnagel E, Baldauf M, Hofmann S and Kataria A 2013 Maritime human factors and IMO policy *Marit. Policy Manag.* **40** 243–60

[10] Knapp S and Van de Velden M 2011 Global ship risk profiles: safety and the marine environment *Transp. Res. Part* D **16** 595–603

[11] Klanac A and Varsta P 2011 Design of marine structures with improved safety for environment *Reliab. Eng. Syst. Saf.* **96** 75–90

[12] Lin W C and Cheng H H 2021 Improving maritime safety through enhancing marine process management: the application of Balanced Scorecard *Manag. Decis.* **59** 604–15

[13] Legato P and Mazza R M 2018 A decision support system for integrated container handling in a transshipment hub *Decis. Support Syst.* **108** 45–56

[14] Vander Hoorn S and Knapp S 2015 A multi-layered risk exposure assessment approach for the shipping industry *Transp. Res. Part* A **78** 21–33

[15] Cheng H H and Ouyang K 2020 Development of a strategic policy for unmanned autonomous ships: a study on Taiwan *Marit. Policy Manag.* **48** 316–30

[16] Banda O A V and Goerlandt F 2018 A STAMP-based approach for designing maritime safety management systems *Saf. Sci.* **109** 109–29

[17] Ahvenjärvi S 2016 The human element and autonomous ships *TransNav: Int. J. Mar. Navig. Safety Sea Transp.* **10** 517–21

[18] Porathe T, Hoem Å, Rødseth Ø J, Fjørtoft K and Johnsen S O 2018 At least as safe as manned shipping? Autonomous shipping, safety and 'human error' *Safety and Reliability—Safe Societies in a Changing World. Proc. of ESREL 2018(June 17–21, 2018) (Trondheim, Norway)*

[19] Evensen M H 2020 Safety and security of autonomous vessels. Based on the Yara Birkeland project *Master's Thesis* The University of Bergen

Chapter 5

Concerns and challenges

MASS sailing is going to become a reality soon. Before long, people may be able to take trips on the open sea with MASS. However, there's a big difference between traveling in a MASS autopilot ship or taking control of an unmanned boat. For example, imagine a situation where a MASS carrying over 20 000+ TEU is about to run into a chemical tanker and a fishing boat. You can remotely control MASS, but you have only a few options: one is crash into a nearby island, save the chemical tanker or the fishing boat but lose MASS, or directly hit the fishing boat and tanker, sacrifice them and save over 20 000+ TEU. What will you do?

The above issues relating to the safety, economics and ethics of MASS are just getting started. While MASS can have lots of positive effects, it may also have many negative ones. For example, criminals may attempt to smuggle illegal bulk cash, illegal immigrants, or drugs by using MASS. Meanwhile, as more and more ships and their operating systems are connected to the Internet more frequently, threats to Internet privacy, cybersecurity and data safety on board ships are also very important. It is predicted that wherever MASS goes, criminals may be all over the place. It still has the risk of being secretly recorded by criminals. Even more, cyber hackers are using an anonymous user name on maritime satellites, and continuously attempting to gain greater control, influence, or manipulate a group of MASS. To sum up, one day, if MASS takes you on a sea vacation or work on board on the open seas, you should view the following issues.

5.1 Technological environment

(1) Lack of the construction of infrastructure
Ports are located on all the continents, and people have been using them for thousands of years. Today, many ports have become popular transport spots, places such as the port of Shanghai, Singapore, and Ningbo-Zhoushan, etc, all offering a convenient environment and a chance to

doi:10.1088/978-0-7503-3151-7ch5

connect to the world. Obviously, a friendly shipping environment must begin with facilities with a solid infrastructure. For example, ships must keep in contact with port authorities when ships approach or enter the port. Particularly, when ships have berthed at the port while the cargo is being processed, ships and the port authorities must still keep in contact with each other. Therefore, it can be seen that the communication facilities are very important tools when it comes to maintaining close contact with the port authorities and the ship.

For MASS, if the main deficiency of the port infrastructure is lack of digitization, intelligence and autonomy in the port facilities, it will undoubtedly hinder the further development prospects of MASS. For example, Sakhi *et al* [1] believe that the development of MASS depends on the key port's performance indicators and other necessary facilities. Perera and Batalden [2] propose that to preserve the integrity of system-level intelligence, adequate consistency between human and system collision avoidance actions is expected. Briefly, from the development of autonomous shipping ports, many ports have to offer more efficient, productive and safe port operations with insufficient automation facilities and communications instead of relying on stevedores to communicate by radio and telephone. Therefore, for the realization of the MASS projects, setting up port facilities with a solid infrastructure and improving the port facilities in a more technologically advanced and autonomous friendly manner is second to none.

(2) Severe weather conditions

There is no doubt that severe weather conditions have a significant effect on autonomous sailing patterns. With climate change, strong hurricanes, typhoons, rising water temperatures, intense winds and torrential rains are becoming the new normal at sea. In fact, severe weather of this severity can also have significant impacts on operators and MASS. They result in the loss of cargo, ships, crews, threatening not only the livelihood of crews but also the stability of the entire maritime supply chain. Fortunately, solutions to severe weather crises do exist for MASS [3]. For example, MASS must have numerous monitoring system devices while traveling, such as cameras, LiDAR, and sonar.

These solutions will go a long way toward solving severe weather crises, but they cannot deal with the root cause of the problem. Taking LiDAR and infrared camera as an example, the relationship between weather conditions and long-range radar is very important. Since LiDAR will reduce the effective distance of measurement under the severest weather conditions; the infrared camera cannot easily survey comprehensively the surrounding of MASS. Gradually, it is inevitable that MASS will experience maritime accidents. In other words, when sailing in severe weather conditions with the inability of these sensors to be working effectively, MASS safety is undergoing severe threats. Hence, if we are to truly tackle severe weather crises for MASS sailing, first, we have to improve the efficiency of that which assists sensor operation.

(3) Spectrum sharing

In a major breakthrough in communication research (CR), scientists have demonstrated high range coverage and extreme accurate satellite communication networks, such as 6G (sixth-generation wireless), the next generation of cellular communications and wireless connectivity. It is also the next giant leap toward both terrestrial and satellite communications. Moreover, a new generation of satellite communication networks has even exhibited the ability to offer higher speed data transfer and minimal latency. If this ability can be utilized in the entire maritime supply chain, it could contribute to massive improvements in MASS development. Scientists are already aware that spectrum sharing does enhance spectral efficiency, but now they also have confirmed that lack of spectrum sharing and much lower internet connection latency are also major obstacles to the development of MASS as well.

To show this, one situation can be imagined where if a MASS container ship is sailing in narrow waters or busy waterways, suddenly, the network signal is slowed down or interrupted, it may very likely cause a MASS container ship to collide or ground. Obviously, the special spectrum range must be determined as a prerequisite for supporting the development MASS. This example is like when a phone call gets disconnected in daily life, it would be very annoying. Moreover, if a MASS loses signal, a maritime accident may probably occur, or even more serious danger. Höyhtyä [4] proposed that using spectrum sharing to obtain more resources is especially useful for non-latency critical data in the shared band.

Satellite communication networks are frequently used in 'transportation', but this is one of the most important times they have been used for MASS projects. The resulting application could potentially stimulate the development of new MASS technologies that would enable much efficiency application of certain maritime industries. Of course, once the spectrum range is sufficient, it will contribute rapid development of application in MASS. Predictably, after MASS or maritime industries obtain sufficient spectrum, such efficiency application could also in turn lead to greatly improved chances of competitiveness affecting maritime-related industries.

(4) Internet privacy

Internet privacy is a complicated issue full of conflicting interests, safety, secrecy, and information confusion. Especially when the internet of MASS came into being, people could suddenly connect with each other from anywhere in the world. In fact, internet privacy of MASS is different from security. Security focuses on malicious behaviors that harm people, while privacy focuses on the protection of information and data.

In order to ensure the internet privacy of MASS and reduce the concerns about data collected, first of all, the standards of network security should be strengthened by all network manufacturers, such as requiring all networks to take protection, measures, and encryption procedure. On the other hand, it is also necessary that the design of network security reduces malicious network attacks [5] and network data theft should be strengthened. In brief, the internal monitoring systems should successfully improve any possible network system infringement. Therefore, under the principle of internet privacy protection, security software upgrade must be insisted on by manufacturers to improve the necessity of protecting privacy. For example, Tam and Jones [6] studied the different cyber risks and web security vulnerabilities about MASS and then put forward relevant analysis. Moreover, Kavallieratos *et al* [5] still think that multiple cyber issues remain open and should be addressed as soon as possible. Thus, it can be seen that no study to date has fully analyzed the cyber architecture of MASS, and no potential cyber threats and cyber-attacks have been identified.

In short, to ensure network privacy safety, reducing attack vulnerabilities and network protection is always the important essential for developing MASS projects.

5.2 Social responsibility

(1) Amendments to international convention

When will MASS be able to sail on the open sea? What restrictions does MASS have before sailing on the open sea? While the idea of MASS sailing on the open sea might sound like anticipation, the IMO international convention actually holds the key to the future of MASS.

In less than 30 years since MASS development, MASS sailing on the open sea will become a major issue for many parts of the world. Indeed, for IMO, there are many tough tasks ahead of them, as it has proven difficult to develop fair conventions regarding autonomous ship usage. On the one hand, they want to pass conventions that protect people's safety and privacy. One the other hand, they don't want to set restrictions that will prevent improvements to this technology. Although the current international conventions are still in the drafting stage for MASS, all MASS are unallowed to sail on the open seas until the international conventions are officially revised and implemented. At the same time, because shipping is unique and international, if relevant international conventions are not amended, the progress of the development of MASS may face challenges for many nations. For example, MASS cannot sail under COLREGs before the convention is amended [7]. Furthermore, Ringbom [8] proposes to elaborate on the distinction between the autonomous ship and the level of manning. Furthermore, he also thinks that this complex regulatory challenge is what IMO has to address as the most important mission. On the face of it, if the restrictions on international conventions are not amended, this will hinder the development of MASS.

Nevertheless, there are some actions about allowing MASS to sail on the open sea in the IMO international convention. For example, in April 2022, the 105th session of IMO's MSC (MSC105) commenced to develop the non-mandatory MASS Code, with a view to passing it in the second half of 2024 as the first stage. Additionally, based on the experienced gained in the application of the non-mandatory MASS Code, a mandatory MASS Code will be envisaged to enter into force in 2028. As a result, IMO has helped ensure MASS sailing on the open sea and remove some of the convention restrictions of developing the world's autonomous shipping.

(2) Determine responsibility

The determination of the most influential factors in maritime accidents, also known as the determination of maritime safety responsibility, is a serious and important global maritime work. Thanks to large parts of science and technology, maritime safety around important achievements such as the human factors or the climate crisis are able to provide a basic means of subsistence across the maritime industries. Although today's science and technology make maritime safety better than ever, the phenomenon itself has always been a significant challenge.

Indeed, one can see the seeds of today's ship safety development being planted more than 3400 BC ago. During that time, of course, besides safety, responsibilities were also gradually known as an important task. Nowadays, numerous studies have described that the concepts of safety, responsibility and autonomy are linked in determining the scope of MASS. In order to completely develop MASS technology, some legal responsibilities and obligations must be resolved. Ordinarily, insurance companies often adapt risk assessments based on the operating conditions of the shipping company, ship age, equipment, experience, etc. For a long time, maritime accidents were mostly caused by human factors, and insurance companies also determine the attribution of liability based on the cause of the accidents. However, the condition of MASS is rather special since in most cases, ships will not be controlled by the crews on board in the future. Therefore, for current or future insurance companies, the work of determining responsibility or claims settlement will become very complicated, and the standards of insurance managing compensation would become very difficult. For example, with the sailing of MASS, the traditional role of the captain may disappear. If a maritime accident happens, related insurance and liability issues will completely subvert the legal issues for maritime accidents at present [7]. Therefore, before MASS officially sails on the open seas, MASS still needs to evaluate a lot of accident identification which also includes more complex situations. It can be seen from this that it is an urgent matter to begin carrying out comprehensive determination of responsibility for MASS.

But, will the clarification of responsibility be just a hot issue that is being discussed now and later be forgotten? Indeed, more and more people working in the maritime industry are learning how to identify maritime safety responsibility. But, one thing is for certain: as long as MASS projects have sustainable development needs, clarification of determining responsibility can only be seen through time.

(3) Ethical anxiety

Today, our world faces a huge number of ethical dilemmas. As issues like conflict between different principles or values have become more serious, it is only natural that people will worry more about the state of the world. Almost everyone has concerns like these, so you are not alone if you sometimes feel upset or hesitate about how humans must choose between behavior. But for some people, these worries can develop into a more serious global problem also called ethical anxiety.

Ethical anxiety is gradually becoming one of the major components of the MASS development process. Indeed, many nations and regions have their own national conditions they make people comply with. However, some people would rather hurt themselves than cause harm to others, even though it may be their only choice. Sometimes, these choices have nothing to do with the requirement of law. Instead, they are related to ethical dilemmas. For example, the situation where a MASS carrying over 20 000+ TEU is about to run over a chemical tanker and a fishing boat. You can remotely control MASS, but you have only some choices: crash into a nearby island, save the chemical tanker or the fishing boat but lose MASS, or hit the tanker and boat directly, damage them but save the over 20 000+ TEU. Some people choose to make decisions based on what many would call moral principles or values. Not surprisingly, what is particularly important in the development of ethical concerns is that people are both permitted to follow the rules of the nations and protect people and property while continuing to follow their autonomous policy. This means that the autonomous policy aspects have to ensure the fullest discussion of all issues required, while still having the individual and social needs.

In fact, these kinds of feelings are understandable. However, it is important not to let negative thoughts affect your choices. Although it is impossible for one person to save the world, an individual can still do many things. For example, you can take personal action to help you manage what is within your control then make the right choice, especially never forgetting your power as an individual. You may be surprised by how much positive change you can promote in your own life and the MASS environment if you really try.

(4) The threat of pirates

By now, everyone knows that pirates are one of the major threats to the maritime industry. However, in our daily lives, this is a threat that we often ignore. The fact is that pirates are already affecting some of our operations in the global maritime industrial supply chain, and will also force us to change our maritime behavior sooner rather than later.

Safety, for example, is an important consideration for autonomous technology. At present, many MASS ships are equipped with high-end sensors, devices, components, and equipment, whose purpose is to maintain the safety of MASS. Among them, one of the major reasons is related to the attack of pirates. However, in order to maintain the security of these sensing devices, information connections, human interference, hacking attacks, and malicious behaviors, the autonomous ship's networking must be protected. In addition, if pirates can hack into the network of a MASS control system, pirates may achieve the purpose of intercepting the ship. Or when a network virus enters the shore side network control system, cyber hackers may disrupt the communication system, causing the autonomous ship to fail to sail as scheduled. For example, Rødseth *et al* [9] think that communication between the ship side and shore control cnter (SCC) is critical for MASS. Thus, adequate and proper safety and security precautions are required. Therefore, in order to avoid similar attacks and improve network safety, network managers or designers also need to carefully examine the ship networking and malicious attacks by hackers.

To keep these malicious attacks from emerging, many governments and their policy decision-makers, researchers and scientists have been working together to offer sustainable policies that provide more resources and a safer environment to change malicious attacks. They have had a lot of success so far, but these policies still cannot completely stop cases of malicious attacks. Therefore, if we do not stop pirates or hackers, we may have to continue to improve the safety and security precautions of MASS to deal with these malicious attacks in the future.

(5) The impact of commercial profit

In recent years, discussion of prioritizing safety of MASS has grown significantly. This collective urge to be constantly safe has become so widespread that it has been constantly discussed: can we reduce crew size to improve safety for MASS sailing? Those caught up in this issue often put a lot of emphasis on it, and to some extent, could be described as addicted to the importance of safety. On the other hand, does 'reduce crew size to improve safety and gain commercial profits for MASS sailing' sound stressful and exhausting, well, it actually is. While some people are in the cycle of working and destroying, some people are convinced that this is a positive thing.

Promotors embracing this philosophy may feel like they're sacrificing crews' careers for a greater maritime good—whether that is the benefit of their company's management or their company's cost. But this kind of myth almost always ends up doing people more harm than good. The author thinks that reducing too much work results in an overly extreme discussion in safety manners which can lead to serious social issues and increased risk of unemployment rate. The reason for this is because the new derived problems of MASS are a huge global maritime issue. In fact, some of the new problems of MASS are derived from problems of existing traditional ships. This is a foreseeable problem in the process of MASS development. Even if these traditional ship's derivative problems could be addressed, it would be enough to improve one of the problems of MASS development on related human issues. However, MASS problems are not only human safety problems, but also an important issue of commercial profit, too.

Clarifying the attribution of responsibility, guarantee, and settlement of insurance claims requires immediate solution as well as a formal, well-defined syntax and semantics. In fact, what does human error mean in the autonomous ship? From another perspective, if commercial profit is a problem that cannot be addressed, it would be an important barrier of autonomous ship development, second only to problems from human error. Clearly, something needs to be done.

The author thinks that solving the problem of commercial profit would require a many-pronged approach. From the insurer and the insured level, crews or ship owners could set a foundation by carefully planning their contracts and creating better insurance contents. Enterprises could work with organizations, fully communicate with the insurance company and the labor union instead of sticking to their own views and saying their own words. Governments can enact legislation to ensure that contracts are provided to insured persons or others. If we all work together, we can guarantee that consensus is supported by those who need it and that our precious discussion results are not wasted.

In conclusion, more and more sensors with advanced functional requirement will be applied in ships to achieve safer, more convenient ships with a higher operational efficiency [10]. In particular, with the development of telecommunications, global location systems, and remote Internet access on a global scale, these will become standard features of MASS in the near future. Moreover, in addition to specific technical requirements, socio-economic, and demographic factors, along with the international environmental system are also very important conditions. Take the particularity and internationality of shipping into account, the development and current situation of shipping is more closely related to the development of international transportation. Therefore, for development of MASS, any changes or adjustments must be linked to the international standards. Briefly, it does not make sense to deviate from international norms when it comes to shipping reforms. Regardless, by following the steps of developing autonomous ship project processes, we look forward to resolving more technology-related limits in decades.

References

[1] Sakhi F E, Ait Allal A, Mansouri K and Qbadou M 2022 Towards determination of key port performance indicators for implementation of autonomous ship *Advanced Intelligent Systems for Sustainable Development (AI2SD'2020)* (Cham: Springer International Publishing) 1 pp 1124–37

[2] Perera L P and Batalden B M 2019 Possible COLREGs failures under digital helmsman of autonomous ships *Oceans 2019-Marseille* (Piscataway, NJ: IEEE) pp 1–7

[3] Jokioinen E, Poikonen J, Jalonen R and Saarni J 2016 Remote and autonomous ships-the next steps *AAWA Position Paper* (London: Rolls Royce PLC)

[4] Höyhtyä M 2019 Connectivity manager: ensuring robust connections for autonomous ships *2019 2nd Int. Conf. on Intelligent Autonomous Systems (ICoIAS)* (Piscataway, NJ: IEEE) pp 86–90

[5] Kavallieratos G, Katsikas S and Gkioulos V 2019 Cyber-attacks against the autonomous ship *Computer Security: ESORICS 2018 Int. Workshops, CyberICPS 2018 and SECPRE 2018, Barcelona, Spain, September 6–7, 2018, Revised Selected Papers 2* (Berlin: Springer International Publishing) pp 20–36

[6] Tam K and Jones K 2018 Cyber-risk assessment for autonomous ships *2018 Int. Conf. on Cyber Security and Protection of Digital Services (Cyber Security)* (Piscataway, NJ: IEEE) pp 1–8

[7] Carey L 2017 *All Hands Off Deck? The Legal Barriers to Autonomous Ships. [ebook]* (Singapore: National University of Singapore)

[8] Ringbom H 2019 Regulating autonomous ships—concepts, challenges and precedents *Ocean Dev. Int. Law* **50** 141–69

[9] Rødseth Ø J, Kvamstad B, Porathe T and Burmeister H C 2013 Communication architecture for an unmanned merchant ship *2013 MTS/IEEE OCEANS-Bergen* (Piscataway, NJ: IEEE) pp 1–9

[10] Ma F, Zhang B, Chen C and Liu J 2018 Design and application of multi-function marine simulators for the future *Int. Conf. on Marine Simulation and Ship Maneuvarability (MARSIM 2018)*

IOP Publishing

Trends in Maritime Autonomous Surface Ships (MASS) Technology

Hsin-Hung Cheng

Chapter 6

Conclusion and future trends

Ships have been developing for more than five thousand years. During this time, they have always been the main mode of transportation, commerce, and trade in the world today. For us, at the beginning of the 21st century, it is believed that we are living in another crucial age in ship history. For example, fully electric ships are set to replace the fuel-consuming ships, and MASS may soon make human crew a thing of the past. In many instances, this is still happening right now.

Autonomous ship technologies will be the main driving force for the future maritime industry. Ship technology has been rapidly developing in the past, but crew size has been declining. For example, in the 18th century, there would have been over one hundred crew members in large ocean-going ships. As ship technology developed, however, crew size declined to less than 18 in a ship of over 20 000+ TEU. Moreover, according to the evidence of related studies, a single crew member or even crewless may become a trend in the future. It is not surprising that automation and autonomous technology will consistently play a key role in maritime industry. Meanwhile, thanks to science and technology, we can already imagine that MASS will sail on the open seas in the future. On the other hand, due to the development of MASS work which has gained ground for the past 10 years, it is predicated that new and unanticipated challenges are waiting for us.

Although MASS has encountered challenges during the development process, it has also brought a whole new opportunity to the maritime industry. For example, as 'Background of artificial intelligence (AI)' in chapter 2, there are many human works that needs to be handled by AI in actual daily life. Fortunately, in these stages, some of the human work has been adequately addressed by AI, which not only simply replaced humans, but also made human work more efficient. In addition, in chapter 3 'Evolution of MASS', while many visual spectrum cameras, LiDAR, or IoV, etc, have been installed on MASS, understandably, these technologies have made considerable progress in the consideration of cost and time.

doi:10.1088/978-0-7503-3151-7ch6

It is worth mentioning that if MASS is successfully to employ full electric propulsion, the ships will have advantages to address like cutting down on shipping costs, reducing marine pollution, etc. On the other hand, by installing AI into the fully autonomous navigation, it is expected to reduce human errors, prevent accidents, as well as alleviate shortage of ship operators. In chapter 4, based on the impact, development and potential for MASS that was mentioned, it is found that many researchers have argued that the commercial adoption of MASS would lead to crew unemployment, or even to eventual disappearance of the crew's profession. However, the majority of people believe that new shore job prospects will arise with the development of MASS, at the same time, crews will gain more MASS-related training and educational courses. For example, besides possessing an information technology (IT) background, the Remote Control Centres (RCCs) must know how the ships wait for a suitable tide or wind to cross the obstacles between the sea, shallow waters, rocky areas, and rivers to avoid grounding. On the other hand, when it comes to maritime industry impact and influence of port transportation, the definition based on its regulatory scope exercise has been widely drawn upon by diverse stakeholders. However, they may not seem to fully recognize the breadth and depth of regulatory scoping exercise, including the formation of non-standard risk related management, codes of conduct, guidelines, and considerations. Therefore, it is necessary for the government of nations to form a comprehensive regulatory framework governing MASS safety and risk, and implement it as soon as possible.

Even though the current development of MASS has encountered many technical and regulatory scoping exercise difficulties, humans have been hard at work to overcome many generated possible problems in recent decades. Nevertheless, problem-solving will be a trend that is very clear in the foreseeable future. More significantly, I expect you to learn a lot of practical skills from this book, and then develop the ability to address many challenges. In any case, at least so far, readers have already understood the many developments of MASS, which of course still require more delicate handling. Therefore, this book has collected much relevant MASS information, hoping that readers can understand the role we play in this era while living in the development of the MASS era.

Thus, what can be predicted is that autonomous technology continues to change the business models of the maritime industry. And this time, this change will not just bring convenience to the ocean carrier, but also offer a whole new opportunity for the global maritime industry. When autonomous technology continues to develop, how can we think positively and actively deal with the problems brought about by autonomous technology, especially to those working hard in the maritime industry? In other words, are we ready to accept everything that MASS brings us? No matter what the answer is, it is safe to say that MASS is all the rage at this moment, and its potential has only just started to begin.

www.ingramcontent.com/pod-product-compliance
Ingram Content Group UK Ltd.
Pitfield, Milton Keynes, MK11 3LW, UK
UKHW051848100926
8556IPUK00005B/511

9 780750 331494